Holt Mathematics

Chapter 1 Resource Book

HOLT, RINEHART AND WINSTON

A Harcourt Education Company

Orlando • Austin • New York • San Diego • London

ISBN 0-03-078296-1

1 2 3 4 5   170   09 08 07 06

CONTENTS

Holt Mathematics

Holt Mathematics

Date ____________

Dear Family,

In this chapter, your child will learn about patterns, exponents, scientific notation, order of operations, and properties of numbers, Your child will work with variables and algebraic expressions, and learn how to solve equations. He or she will also work with metric units of measurement.

Your child will identify and extend numeric and geometric patterns.

The pattern is to add 1 more than the time before.   The next three numbers are:

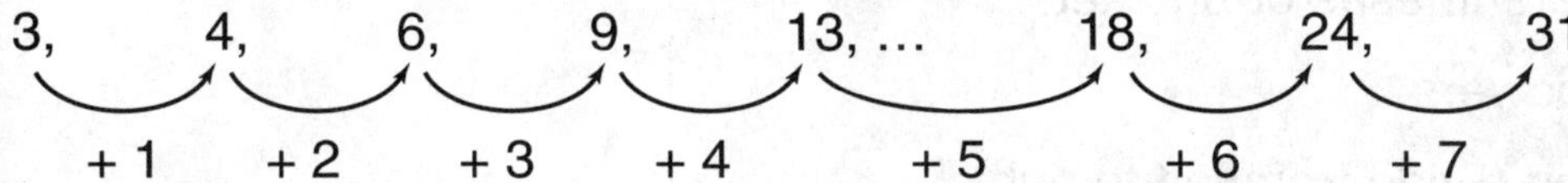

Using **exponents** is an efficient way of expressing large numbers. Exponents typically appear in scientific work and publications.

An exponent tells how many times to multiply a number, the base, by itself.

$$\text{base} \longrightarrow 5^2 \longleftarrow \text{exponent}$$

5^2 expresses 5×5. The total value equals 25.

When dealing with very large numbers, **scientific notation,** a kind of shorthand, is used. In order to write a number in scientific notation:

1. Move the decimal point to create a number between 1 and 10.

2. Count the number of places you moved the decimal to create that number.

3. The number of places you moved the decimal is the power of 10 for that number.

In scientific notation the number 17,900,000 is expressed as:

$$1.79 \times 10^7$$

This is an **algebraic expression.**

$$\text{variable} \longrightarrow n + 3 \longleftarrow \text{constant}$$

Holt Mathematics

Your child will learn to find the value of an expression by replacing the variable with a number and using that number to compute the result.

Find the value of $(n + 3) - 10$ when $n = 15$.

$(15 + 3) - 10$
$18 \quad - 10 = 8$

When simplifying expressions, we use **the order of operations**, a special set of rules to follow when there is more than one operation in the equation.

> **1.** Perform operations within grouping symbols, such as parentheses or brackets.
>
> **2.** Evaluate powers.
>
> **3.** Multiply and divide from left to right.
>
> **4.** Add and subtract from left to right.

Simplify: $4^2 \cdot (3 + 7) \div 2$

$4^2 \cdot (3 + 7) \div 2 = 4^2 \cdot 10 \div 2$	Add within parentheses.
$= 16 \cdot 10 \div 2$	Evaluate power.
$= 160 \div 2$	Multipy.
$= 80$	Divide.

We also use properties of numbers to simplify expressions.

Commutative Property	$a + b = b + a \qquad a \cdot b = b \cdot a$
Associative Property	$a + (b + c) = (a + b) + c$ $a \cdot (b \cdot c) = (a \cdot b) \cdot c$
Identity Property	$a + 0 = a \qquad a \cdot 1 = a$
Distributive Property	$a \cdot (b + c) = (a \cdot b) + (a \cdot c)$

To solve equations, children use **inverse** or opposite operations. Addition and subtraction are inverse operations. Multiplication and division are also inverse operations.

Solve: $n + 7 = 11$

1. In the above equation, n is added to 7.
2. In order to solve for n, you subtract 7 from both sides.
$n + 7 - 7 = 11 - 7$
3. $n = 4$

For additional resources, visit go.hrw.com and enter keyword MS7 Parent.

Holt Mathematics

Name _________________________________ Date __________ Class __________

Practice A
Numbers and Patterns

Write a rule that describes the pattern. Then write the next three numbers in the pattern.

1. 7, 10, 13, 16, _____, _____, _____, ...

The pattern is

____

____

2. 81, 70, 59, 48, _____, _____, _____, ...

The pattern is

____

____

3. 2, 4, 8, 16, _____, _____, _____, ...

The pattern is

____

____

4. 1, 3, 6, 10, 15, _____, _____, _____, ...

The pattern is

____

____

Write a rule that describes the pattern. Then draw the next three figures in the pattern.

5. 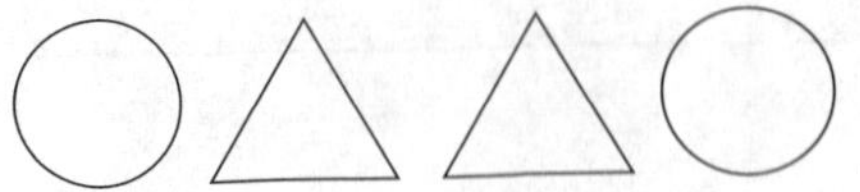

The pattern is _______________________________

6. 

The pattern is _______________________________

7. Complete the table so that it shows the number of dots in each figure.

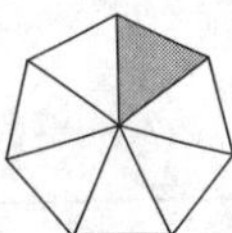 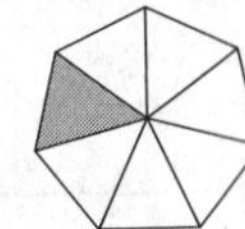 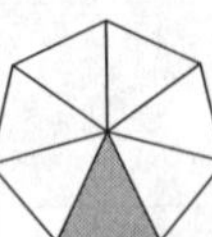 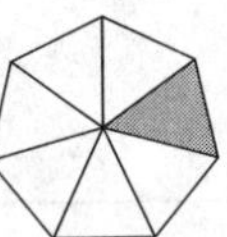

Figure 1   Figure 2   Figure 3   Figure 4   Figure 5

Figure	1	2	3
Number of Dots			

How many dots are in the fifth figure? _____

Use drawings to justify your answer.

3

Holt Mathematics

Name ____ Date __________ Class __________

Practice B
Numbers and Patterns

Identify a possible pattern. Use the pattern to write the next three numbers.

1. 41, 37, 33, 29, _____, _____, _____, ...

____

____

2. 50, 52, 56, 62, _____, _____, _____, ...

____

____

3. 320, 160, 80, 40, _____, _____, _____, ...

____

____

4. 24, 40, 56, 72, _____, _____, _____, ...

____

____

Identify a possible pattern. Use the pattern to draw the next three figures.

5.

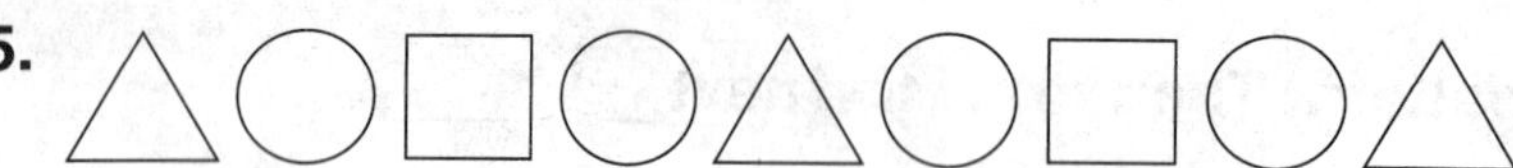

____

6.

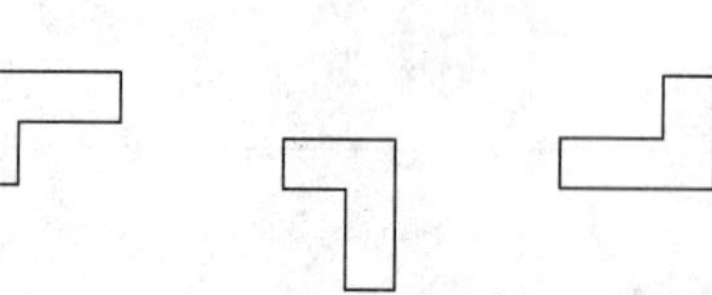

____

7. Complete the table so that it shows the number of dots in each figure.

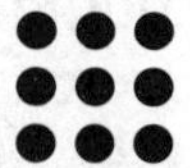

Figure 1   Figure 2   Figure 3   Figure 4   Figure 5

Figure	1	2	3
Number of Dots			

How many dots are in the fifth figure of the pattern? _____

Use drawings to justify your answer.

4

Holt Mathematics

Practice C
Numbers and Patterns

Identify a possible pattern. Use the pattern to write the missing numbers.

1. 72, 95, 118, 141, _____, _____, _____, ...

2. 65, _____, 51, 44, _____, _____, 23, ...

3. 4, 12, _____, 108, 324, 972, _____, ...

4. _____, _____, 47, 44, 40, _____, 29, ...

Look for a possible pattern. Use the pattern to draw the next three figures.

5. 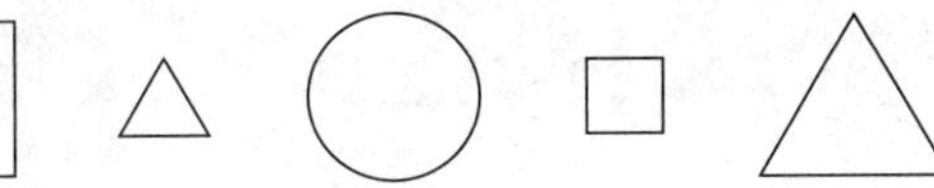

6. 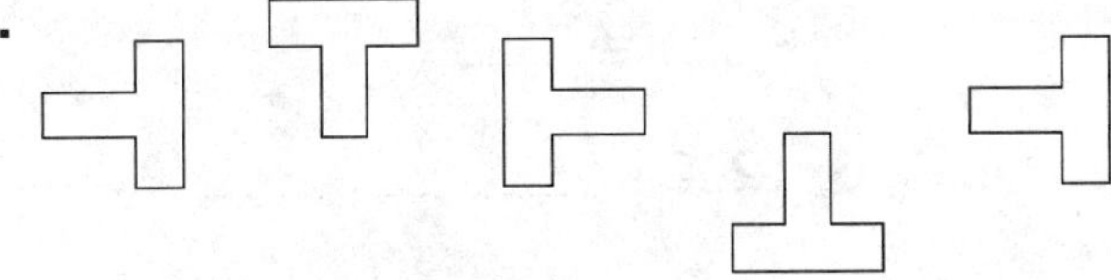

7. 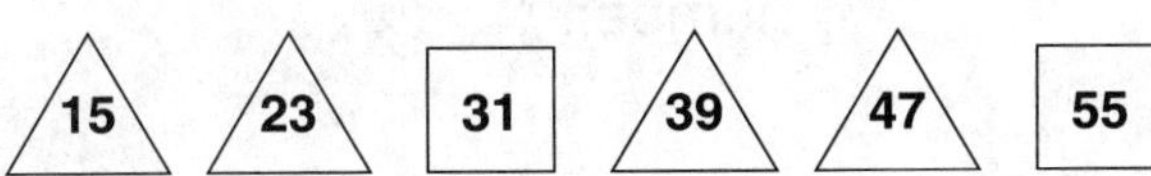

8. Complete the table so that it shows the number of squares in each figure.

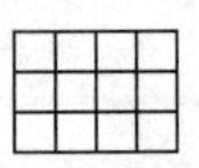

Figure 1   Figure 2   Figure 3   Figure 4   Figure 5

Figure	1	2	3
Number of Squares			

How many squares are in the fifth figure of the pattern? _____

Use drawings to justify your answer.

Holt Mathematics

Name ____ Date _________ Class _________

Reteach
Numbers and Patterns

To identify a number pattern, ask: What can I do to each number to get the number that comes next?

The pattern is to add 4 to get the next number. Use the pattern to get the next three numbers.

2   6   10   14   □   □   □

$2 + 4 = 6$ $6 + 4 = 10$ $10 + 4 = 14$ $14 + 4 = 18$ $18 + 4 = 22$ $22 + 4 = 26$

and   but

$2 \times 3 = 6$ $6 \times 3 \neq 10$

Identify the pattern. Use the pattern to find the next three numbers.

1. 28, 25, 22, ___, ___, ___, ...

Subtract ___ to get the next number.

$22 - \underline{} = \underline{}$

$\underline{} - \underline{} = \underline{}$

$\underline{} - \underline{} = \underline{}$

2. 6, 24, 42, 60, ___, ___, ___, ...

Add ___ to get the next number.

$60 + \underline{} = \underline{}$

$\underline{} + \underline{} = \underline{}$

$\underline{} + \underline{} = \underline{}$

3. 4, 12, 20, 28, ___, ___, ___, ...

___________ to get the next number.

4. 3, 6, 12, 24, ___, ___, ___, ...

___________ to get the next number.

To identify a geometric pattern, ask: How can I change each figure to get the next figure?

The pattern is to shade the next square in a clockwise direction. Use the pattern

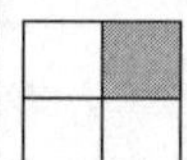 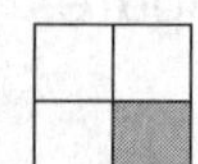 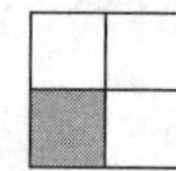 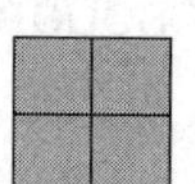 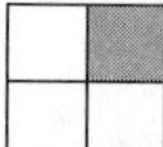   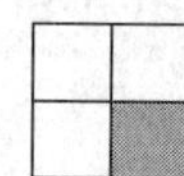 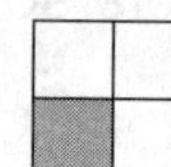 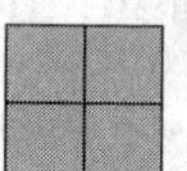

Draw the next three figures in each pattern.

5. 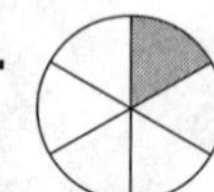 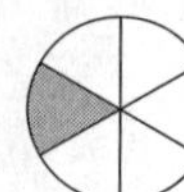 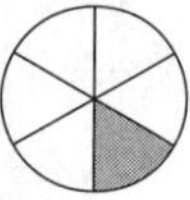 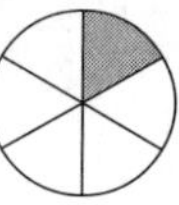 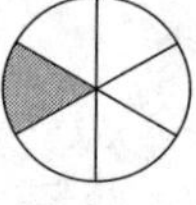

6. 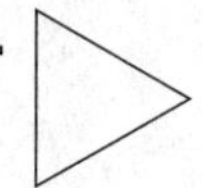  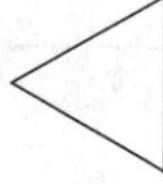 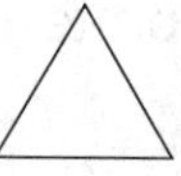 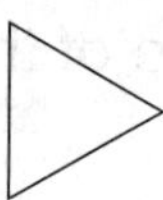

Holt Mathematics

LESSON 1-1 Challenge
Pascal's Triangle

The pattern at the right is called Pascal's Triangle. This pattern is named after the French mathematician, Blaise Pascal, because he wrote about its properties.

Use Pascal's Triangle to solve problems 1-5.

Row 0	1
Row 1	1 1
Row 2	1 2 1
Row 3	1 3 3 1
Row 4	1 4 6 4 1
Row 5	1 5 10 10 5 1
Row 6	? ? ? ? ? ? ?
Row 7	? ? ? ? ? ? ? ?
Row 8	? ? ? ? ? ? ? ? ?
Row 9	? ? ? ? ? ? ? ? ? ?

1. Write the sum of the numbers in each of rows 1-5. What pattern do you see?

____

____

2. Use the pattern that you found in problem 1 to predict the sum for row 9. _________

3. Look at the numbers in row 4 and the numbers they are connected to in row 3. Then look at the numbers in row 5 and the numbers they are connected to in row 4. What pattern do you see?

____

____

____

4. Use the pattern you found in problem 3 to find the numbers for rows 6-9.

   row 6: _______________________________________

   row 7: _______________________________________

   row 8: _______________________________________

   row 9: _______________________________________

5. Find the sum of the numbers in row 9. How does it compare to your prediction in problem 2? _______________________________________

Holt Mathematics

Name _________________________________ Date __________ Class __________

 Problem Solving
Numbers and Patterns

Write the correct answer.

1. Trains leave Peapack station at 6:40 A.M., 7:16 A.M., 7:32 A.M., and 7:48 A.M. Assume that the pattern continues. If you arrive at the station at 8:30 A.M., at what time will the next scheduled train leave?

2. Suppose the pattern 8, 16, 24, 32, ... is continued forever. Will the number 174 appear in the pattern? Why or why not?

3. A water tank holding 100 gallons begins to leak at 6:00 P.M. At 7:00 P.M. the tank has 94 gallons. At 8:00 P.M the tank has 88 gallons. At 9:00 P.M. the tank has 82 gallons. If the pattern continues, how many gallons will be left at 11:00 P.M.?

4. Awilda is making a necklace with sphere, pyramid, and cube shaped beads. If she continues the pattern below, what are the shapes of the next five beads?

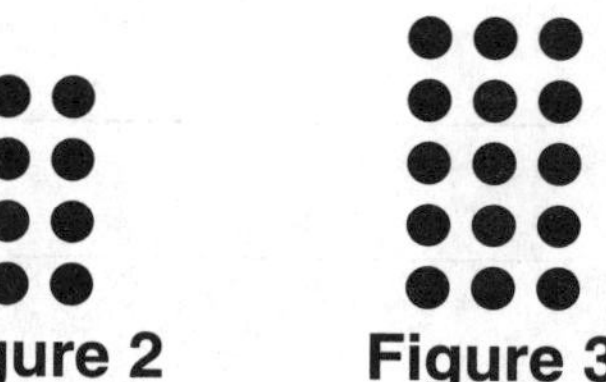

Choose the letter for the best answer.

5. The drawing at the right shows the first three figures in a pattern. If the pattern continues, how many circles will be in the fifth figure of the pattern?

   A 20 C 27

   B 24 D 35

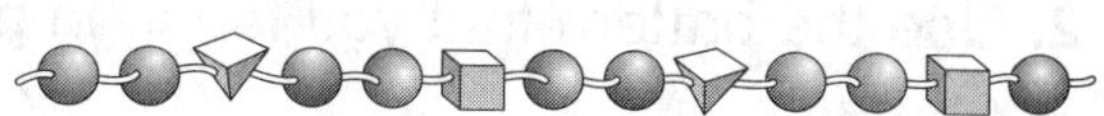

Figure 1   Figure 2   Figure 3

6. The table shows the number of handshakes if each person in a group shakes each other's hand once. How many handshakes will there be if there are 9 people?

   F 20 H 36

   G 30 J 45

People	2	3	4	5	6
Number of Handshakes	1	3	6	10	15

7. Melissa wrote the following number pattern: 71, 66, 59, 50, 39, ... What is the next number in the pattern?

   A 26 C 30

   B 27 D 35

8. By July 1, Bob saved $120. By July 8, he saved $185. By July 15, Bob saved $250. If the pattern continues, how much will he have saved by July 29?

   F $315   H $390

8

Holt Mathematics

Reading Strategies
Identify Relationships

To identify and extend a **number pattern**, you must find the
relationship between each number in the pattern and the number
that it comes after.

4, 16, 28, 40, □, □, □, ...

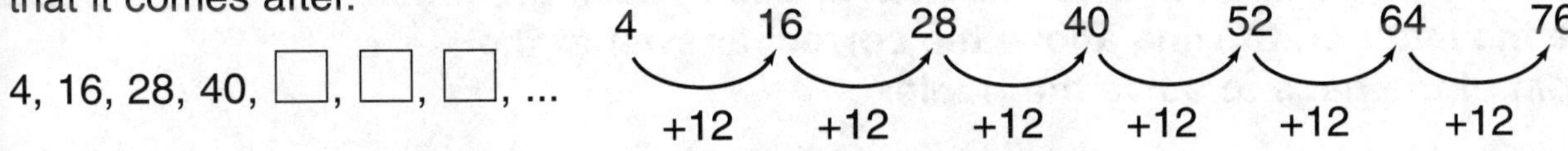

The relationship is that each number is 12 more than the number it
comes after. So, the pattern is to add 12 to each number to get the
next number.

To identify and extend a **geometric pattern**, you must find the
relationship between each figure in the pattern and the figure that it
comes after.

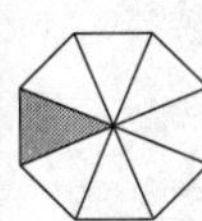 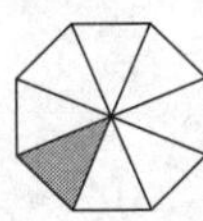 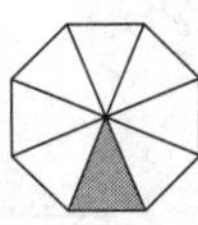 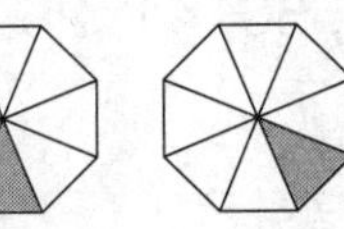   ?   ?   ?

The relationship is that the shaded triangle moves one triangle
counterclockwise from the figure it comes after.

So, the next three figures in the pattern are:

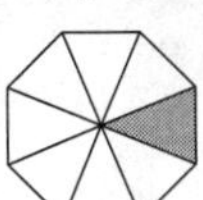 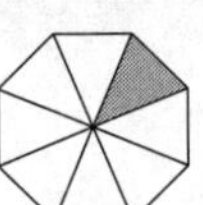 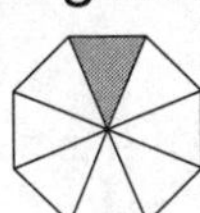

**Use the pattern 64, 55, 46, 37, □, □, □, ... to answer
questions 1 and 2.**

1. What is the relationship between each number and the number it comes after?

   ____

2. What are the next three numbers in the pattern?

   ____

Use the pattern below to answer questions 3 and 4.

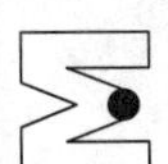  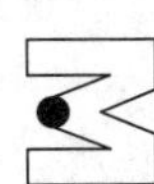  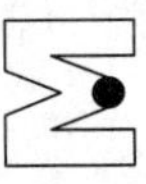

3. What is the relationship between each figure and the figure it comes after?

   ____

Holt Mathematics

Puzzles, Twisters & Teasers

LESSON 1-1

It's in the Cards!

When is it dangerous to play cards?

Write the next number or draw the next shape in each pattern.
Write the letter on the line above the correct answer at the
bottom of the page to solve the riddle.

T   75, 59, 43, 37, _____, ...

E   7, 14, 28, 56, _____, ...

H   12, 25, 39, 54, _____, ...

J   243, 81, 27, 9, _____, ...

W   28, 34, 42, 52, _____, ...

K   114, 91, 68, 45, _____, ...

R   2, 8, 32, 128, _____, ...

N   70, 56, 41, 25, _____, ...

S   5, 8, 14, 23, _____, ...

O   6, 30, 54, 78, _____, ...

L   6, 18, 54, 162, _____, ...

I   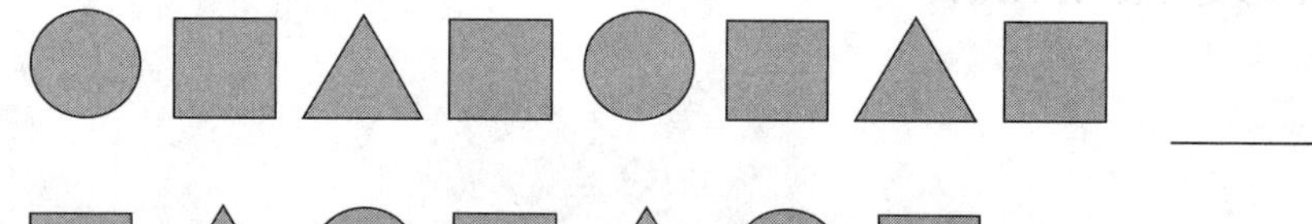

D

___ ___ ___ ___   ___ ___ ___
64   70   112   8   21   70   112

___ ___ ___ ___ ___   ___   ___   ___   ___
3   102   22   112   512   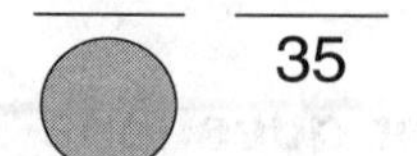   35   64   486

Holt Mathematics

Practice A
Exponents

Multiply.

1. 4^2

$4 \cdot 4 =$ _______

2. 2^3

$\_ \cdot \_ \cdot \_ =$ _______

3. 6^2

$\_ \cdot \_ =$ _______

4. 9^2

5. 4^3

6. 3^5

7. 7^0

____

8. 10^2

9. 3^4

10. 9^1

11. 2^5

____

Use an exponent and the given base to write each number.

12. 25, base 5

13. 3, base 3

14. 8, base 2

15. 1, base 4

16. 81, base 9

17. 64, base 4

18. 64, base 8

19. 9, base 3

20. 36, base 6

21. 16, base 2

22. 27, base 3

23. 400, base 20

24. The first day, Jessie has $2. The second day, she has twice as much money as the first day. The third day, she has twice as much money as the second day. Write the amount of money she has on the third day in exponential form. Then write the amount in standard form.

____

25. Kevin runs 5 miles on Monday. The total number of miles he runs that week is 5 times the number of miles he runs on Monday. How many miles does Kevin run that week?

____

Holt Mathematics

LESSON 1-2 Practice B
Exponents

Find each value.

1. 5^2 **2.** 2^4 **3.** 3^3 **4.** 7^2

________   ________   ________   ________

5. 4^4 **6.** 12^2 **7.** 10^3 **8.** 11^1

________   ________   ________   ________

9. 1^6 **10.** 20^2 **11.** 6^3 **12.** 7^3

________   ________   ________   ________

Write each number using an exponent and the given base.

13. 16, base 4   **14.** 25, base 25   **15.** 100, base 10   **16.** 125, base 5

________   ________   ________   ________

17. 32, base 2   **18.** 243, base 3   **19.** 900, base 30   **20.** 121, base 11

________   ________   ________   ________

21. 3,600, base 60   **22.** 256, base 4   **23.** 512, base 8   **24.** 196, base 14

________   ________   ________   ________

25. Damon has 4 times as many stamps as Julia. Julia has 4 times as many stamps as Claire. Claire has 4 stamps. Write the number of stamps Damon has in both exponential form and standard form.

____

26. Holly starts a jump rope exercise program. She jumps rope for 3 minutes the first week. In the second week, she triples the time she jumps. In the third week, she triples the time of the second week, and in the fourth week, she triples the time of the third week. How many minutes does she jump rope during the fourth week?

____

Holt Mathematics

 Practice C
1-2 *Exponents*

Find each value.

1. 3^8 **2.** 7^5 **3.** 50^2

___________   ___________   ___________

4. 9^4 **5.** 30^3 **6.** 12^5

___________   ___________   ___________

Compare. Write $<$, $>$, or $=$.

7. 7^2 ☐ 48   **8.** 9^3 ☐ 810   **9.** 6^4 ☐ 10^3

10. 100,000 ☐ 10^6 **11.** 2^7 ☐ 6^3 **12.** 4^6 ☐ 8^4

Write each number using an exponent and the given base.

13. 343, base 7   **14.** 625, base 5   **15.** 1,728, base 12

___________   ___________   ___________

16. 225, base 15   **17.** 1,000,000, base 100   **18.** 1,225, base 35

___________   ___________   ___________

19. 6,561, base 9   **20.** 2,187, base 3   **21.** 8,000, base 20

___________   ___________   ___________

22. Jacob has 7 times as many postcards as Austin has. Austin has
7 times as many postcards as Angela has. Angela has 7 times
as many postcards as Samuel has. Samuel has 7 postcards.
Write the number of postcards Jacob has in both exponential
form and standard form.

____

Holt Mathematics

LESSON 1-2 · Reteach
Exponents

The exponent tells you how many times to multiply the base
by itself.

base ⟶ 3^4 ⟵ exponent

- To find 3^4, multiply the base (3) times itself 4 times.

$$3^4 = 3 \cdot 3 \cdot 3 \cdot 3 = 81$$

Find each value.

1. $4^3 =$ _____ $\cdot$ _____ $\cdot$ _____ $= 64$

2. $1^5 =$ _____ $\cdot$ _____ $\cdot$ _____ $\cdot$ _____ $\cdot$ _____ $=$ _____

3. 5^2 **4.** 2^3 **5.** 3^3 **6.** 6^2

__________   __________   __________   __________

7. 8^2 **8.** 4^1 **9.** 5^3 **10.** 2^4

__________   __________   __________   __________

- You can write 64 using an exponent with the base 8.

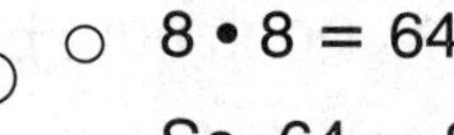

$8 \cdot 8 = 64$

So, $64 = 8^2$.

Write each number using an exponent and the given base.

11. 216, base 6: $=$ _______ and $36 \cdot 6 =$ _________ so, $216 =$ _______.

12. 16, base 4   **13.** 8, base 2   **14.** 9, base 3

__________   __________   __________

15. 81, base 9   **16.** 27, base 3   **17.** 49, base 7

__________   __________   __________

Holt Mathematics

Challenge
1-2 *Money Grows*

If an investment of $50 doubles in value, your investment will be worth $50 • 2 = $100.

If it doubles again, your investment will be worth $50 • 2 • 2 = $200. You can also write this as $50 • 2^2 = $200.

If your money doubles a third time, it will be worth $50 • 2^3 = $400.

Write the correct answer.

1. If you invest $400, how much will it be worth if it doubles in value twice?

2. If an investment of $550 doubles in value twice, how much will it be worth?

3. If you invest $75, how much will it be worth if it doubles in value 4 times?

4. If you invest $125, how much will it be worth if it doubles in value 4 times?

5. If an investment of $75 triples in value 3 times, how much will it be worth?

6. If an investment of $125 triples in value 4 times, how much will it be worth?

When Jasmine turned 13, she started saving $5 per month until she turned 18. She then invested her total savings so that it doubled in value every 7 years.

7. How much money did Jasmine save by the time she turned 18? _______________

8. How many times will her investment double from age 18 to age 60? _______________

9. How much will Jasmine's investment be worth when she turns 60? _______________

When Jake turned 13, he started saving $1.50 per week until he turned 19. He then invested his total savings so that it tripled in value every 12 years.

10. How much money did Jake save by the time he turned 19? _______________

11. How many times will his investment triple from age 19 to age 55? _______________

12. How much will Jake's investment be worth when he turns 55? _______________

Holt Mathematics

LESSON 1-2 Problem Solving
Exponents

Write the correct answer.

1. The cells of the bacteria *E. coli* can double every 20 minutes. If you begin with a single cell, how many cells can there be after 4 hours?

2. The population of metropolitan Orlando, Florida, has doubled about every 16 years since 1960. In 2000, the population was 1,644,561. At this doubling rate, what could the population be in 2048?

3. A prizewinner can choose Prize A, $2,000 per year for 15 years, or Prize B, 3 cents the first year, with the amount tripling each year through the fifteenth year. Which prize is more valuable? How much is it worth?

4. Maria had triplets. Each of her 3 children had triplets. If the pattern continued for 2 more generations, how many great-great-grandchildren would Maria have?

Choose the letter for the best answer.

5. A theory states that the CPU clock speed in a computer doubles every 18 months. If the clock speed was 33 MHz in 1991, how can you use exponents to find out how fast the clock speed is after doubling 3 times?

A 33^3

B $3^2 \cdot 33$

C $2^3 \cdot 33$

D 33^2

6. The classroom is a square with a side length of 13 feet and an area of 169 square feet. How can you write the area in exponential form?

F 2^{13}

G 13^2

H 3^{13}

J 13^3

7. In 2000, Wake County, North Carolina, had a population of 610,284. This is about twice the population in 1980. If the county grows at the same rate every 20 years, what will its population be in 2040?

A 915,426

B 1,220,568

C 1,830,852

D 2,441,136

8. The number of cells of a certain type of bacteria doubles every 45 minutes. If you begin with a single cell, how many cells could there be after 6 hours?

F 64

G 256

H 360

J 540

Holt Mathematics

Name _______________________________ Date __________ Class __________

<table><tr><td>LESSON
1-2</td><td>

Reading Strategies
Reading and Understanding Symbols
</td></tr></table>

Exponents are an efficient way to express repeated multiplication.

3^5 —→ is read "3 to the fifth power."

3^5 means **3 is a factor 5 times**: $3 \times 3 \times 3 \times 3 \times 3$.

$3^5 = 243$ —→ is read "3 to the fifth power equals 243,"

or, "The value of 3 to the fifth power is 243."

The **base** names the factor. The **exponent** tells how many times to repeat the base as a factor.

$$\text{base} \rightarrow 3^5 \leftarrow \text{exponent}$$

You can describe some numbers as the product of a repeated factor.

$$\rightarrow 64 = 4 \times 4 \times 4$$

Then write the product with a base and an exponent.

$$\rightarrow 64 = \quad 4^3$$

Answer each question.

1. How do you read 5^3? ______________________________

2. What does 5^3 mean? ______________________________

3. What is the value of 5^3? ______________________________

4. Write about the difference between 5^3 and 3^5.

____

____

5. Write 32 using 2 as the repeated factor.

____

6. Write 32 using an exponent and a base of 2. ______________________________

7. Write 216 using 6 as the repeated factor.

____

8. Write 216 using an exponent and a base of 6. ______________________________

Holt Mathematics

LESSON 1-2 — Puzzles, Twisters & Teasers
Explore Your Power Base!

Why was the computer tired when it got home?

To find out, solve each problem. Write the letter on the line above the correct answer at the bottom of the page. Each answer must match exactly. Some letters will not be used.

E $\quad 3^4 =$ _______________

T $\quad 49,\ \text{base } 7 =$ _______________

K $\quad 10^3 =$ _______________

B $\quad 81,\ \text{base } 9 =$ _______________

I $\quad 2^6 =$ _______________

D $\quad 5^2 =$ _______________

W $\quad 18^1 =$ _______________

C $\quad 25,\ \text{base } 5 =$ _______________

A $\quad 27,\ \text{base } 3 =$ _______________

H $\quad 14^0 =$ _______________

V $\quad 12^2 =$ _______________

R $\quad 256,\ \text{base } 4 =$ _______________

O $\quad 216,\ \text{base } 6 =$ _______________

U $\quad 5^3 =$ _______________

S $\quad 6^2 =$ _______________

9^2	81	5^2	3^3	125	36	81	64	7^2	1	3^3	25

3^3	1	3^3	4^4	25	25	4^4	64	144	81

Holt Mathematics

LESSON 1-3 Practice A
Metric Measurements

Choose the letter for the best answer.

1. The height of a flagpole
   - **A** millimeters
   - **B** centimeters
   - **C** meters
   - **D** kilometers

2. The mass of a bicycle
   - **F** milligrams
   - **G** kilograms
   - **H** grams
   - **J** centimeters

3. The capacity of a spoon
   - **A** liters
   - **B** milliliters
   - **C** kiloliters
   - **D** millimeters

4. The thickness of a piece of cardboard
   - **F** millimeters
   - **G** centimeters
   - **H** kilometers
   - **J** meters

5. The mass of a banana
   - **A** kilograms
   - **B** grams
   - **C** liters
   - **D** milligrams

6. The capacity of a swimming pool
   - **F** kilograms
   - **G** milliliters
   - **H** liters
   - **J** kiloliters

Choose the letter of the best answer.

7. 320 cm = ☐ m
   - **A** 0.032
   - **B** 0.32
   - **C** 3.2
   - **D** 32

8. 2.5 kg = ☐ g
   - **F** 0.0025
   - **G** 0.25
   - **H** 250
   - **J** 2,500

9. 5,600 mL = ☐ L
   - **A** 5,600,000
   - **B** 56,000
   - **C** 56
   - **D** 5.6

10. 780 m = ☐ km
   - **F** 780,000
   - **G** 78,000
   - **H** 0.78
   - **J** 0.078

11. 0.35 g = ☐ mg
   - **A** 0.00035
   - **B** 0.0035
   - **C** 350
   - **D** 3,500

12. 4.3 mm = ☐ cm
   - **F** 0.043
   - **G** 0.43
   - **H** 430
   - **J** 4,300

13. Benjamin is 165 cm tall. His sister is 1.8 m tall. Who is shorter? Explain your answer.

____

14. Jan has a bag of grapes that has a mass of 0.9 kg. Terri has a bag of grapes that has a mass of 830 g. Whose grapes have the greater mass? Explain your answer.

____

Holt Mathematics

Practice B

LESSON 1-3 *Metric Measurements*

Choose the most appropriate metric unit for each measurement. Justify your answer.

1. The capacity of a paper cup

2. The mass of a small poodle.

3. The width of a computer screen

4. The mass of a pencil

Convert each measure.

5. 496 mm to centimeters

6. 0.68 kg to grams

7. 3,800 mL to liters

8. 832 mg to grams

9. 76 km to meters

10. 2.9 cm to meters

11. 0.041kL to liters

12. 14.9 g to milligrams

13. 7,800 cm to meters

14. Sam's laptop computer has a mass of 4.2 kg. Fred's laptop computer has a mass of 4,940 grams. Which computer has the lesser mass? Explain your answer.

15. Elise makes a poster that is 1.5 m tall. Meg makes a poster that is 96 cm tall. Who makes a taller poster? Explain your answer.

20

Holt Mathematics

Practice C
Metric Measurements

Choose the most appropriate metric unit for each measurement. Justify your answer.

1. The thickness of a piece of spaghetti

2. The capacity of a small sink

3. The mass of an envelope

4. The capacity of a gas truck

Convert each measure.

5. 3.06 kL to liters

6. 75 mm to centimeters

7. 604 g to kilograms

8. 0.95 m to centimeters

9. 5.6 L to milliliters

10. 48 mg to grams

Compare. Write >, <, or = for ☐.

11. 0.5 cm ☐ 5 mm

12. 4.6 kg ☐ 462 g

13. 0.03 kL ☐ 200 L

14. 53 m ☐ 5,400 mm

15. 8,840 mL ☐ 9 L

16. 416 mg ☐ 0.4 g

17. Eleanor ran 3.5 km on Friday and 2.8 km on Saturday. Monroe ran 6,400 m on the same two days. Who runs a greater distance? Explain.

18. Tunnel A is 1.6 km long. Tunnel B is 740 meters long. Which tunnel is longer? How much longer?

Holt Mathematics

LESSON 1-3 Reteach
Metric Measurements

You can use the relationships in this table to help you convert metric units.

Length	Mass	Capacity
1 cm = 10 mm 1 m = 100 cm = 1,000 mm 1 km = 100 m	1 g = 1,000 mg 1 kg = 1,000 g	1 L = 1,000 mL 1 kL = 1,000 L

To change larger units to smaller units, multiply.

$$5.3 \text{ L} = \boxed{} \text{ mL}$$
$$5.3 \times 1,000 = 5,300$$
$$5.3 \text{ L} = 5,300 \text{ mL}$$

1 L = 1,000 mL
Liters are larger than milliliters, so multiply by 1,000.
Move the decimal point 3 places to the right: 5.300

To change smaller units to larger units, divide.

$$46 \text{ cm} = \boxed{} \text{ m}$$
$$46 \div 100 = 0.46$$
$$46 \text{ cm} = 0.46 \text{ m}$$

1 m = 100 cm
Centimeters are smaller than meters, so divide by 100.
Move the decimal point 2 places to the left: 046

Convert each measure.

1. 15 cm to millimeters

Think: 1 cm = _________ mm

Multiply 15 by _________.

15 cm = _________ mm

2. 964 g to kilograms

Think 1 kg = _________ g

Divide 964 by _________

964 g = _________ kg

3. 4.9 kL to liters

_________ 4.9 by _________

4.9 kL = _________ mm

4. 3,531 cm to meters

_________ 3,531 by _________

3,531 cm = _________ m

5. 468 mg to grams

468 mg = _________ g

6. 0.69 kL to liters

0.69 kL = _________ L

Holt Mathematics

Challenge
LESSON 1-3

Have a Ball!

The most commonly used metric units of mass are **milligrams (mg)**, **grams (g)**, and **kilograms (kg)**. However, there are other metric units of mass also. These units include **centigrams (cg)**, **decigrams (dg)**, **decagrams (dag)**, and **hectograms (hg)**. The table below shows the equivalent number of grams for each of these units.

Unit	1 kg	1 hg	1 dag	1 g	1 dg	1 cg	1 mg
Number of Grams	1,000 g	100 g	10 g	1 g	0.1 g	0.01 g	0.001 g

The table below shows the masses of balls that are used in different sports. Rewrite the table so that it shows the balls and their masses in order from greatest mass to least mass. Use the same unit to show each mass.

Ball	Mass
Baseball	14.88 dag
Basketball	0.65 kg
Football	4,252 dg
Paddleball	6,502 cg
Ping pong ball	2,450 mg
Softball	198.4 g
Soccer ball	4.536 hg
Tennis Ball	750 dg
Volleyball	28,000 cg

Ball	Mass

Holt Mathematics

LESSON 1-3 **Problem Solving**
Metric Measurements

Write the correct answer.

1. A porcupine has a mass of 27 kg. A cat has a mass of 6,300 g. Which animal has the greater mass?

2. A faucet drips at the rate of 50 L per day. At this rate, how many days will it take for the faucet to drip a total of 1 kL?

3. The distance from Midwood Library to Midwood Middle School is 0.845 km. The distance from Midwood High School to Midwood Library is 872 m. Which school is closer to the library? How much closer?

4. A mouse has a mass of 18 g. A guinea pig has a mass of 0.85 kg. Is the mass of the two animals together greater or less than 1 kg? Explain your answer.

Choose the letter for the best answer.

The table shows the lengths and masses of some members of the cat family.

Members of the Cat Family		
Name	**Length**	**Mass**
Bobcat	864 mm	18,200 g
Jaguar	1.52 m	55 kg
Canada Lynx	91.4 cm	15,900 g
Mountain Lion	1.35 m	85 kg

5. Which cat has the greatest length?
  A bobcat
  B jaguar
  C Canada lynx
  D mountain lion

6. Which cat has the least mass?
  F bobcat
  G jaguar
  H Canada lynx
  J mountain lion

7. What is the difference in mass between the cat with the greatest mass and the cat with the least mass?
  A 66.8 kg
  B 69.1 kg
  C 7,400 g
  D 15,815 g

8. For which two cats is the difference in length closest to 0.5 m?
  F bobcat and mountain lion
  G bobcat and Canada lynx
  H jaguar and mountain lion
  J jaguar and Canada lynx

Holt Mathematics

Reading Strategies

LESSON 1-3

Follow a Procedure

To convert metric units, follow these steps.

653 mg = ☐ g

Step 1: Determine if you are converting to a larger unit or a smaller unit.

1 mg < 1 g
You are converting to a larger unit.

Step 2: Choose the operation that you need to use to convert.
- To convert to a smaller unit, you need to multiply.
- To convert to a larger unit, you need to divide.

Since you are converting to a larger unit, you need to divide.

Step 3: Identify the number by which you must multiply or divide. That number is the number of smaller units that equal 1 larger unit.

1,000 mg = 1 g
So, you need to divide 653 mg by 1,000 to get the number of grams.

Step 4: Calculate the answer.

$653 \div 1,000 = 0.653$
653 mg = 0.653g

Answer these questions for 3.9 m = ☐ cm.

1. Are you converting to a larger unit or a smaller unit? _____________

2. What operation will you use to convert? _____________

3. By what number do you need to multiply or divide? _____________

4. What is the answer to the problem? _____________

Answer these questions for 14.8 L = ☐ kL.

5. Are you converting to a larger unit or a smaller unit? _____________

6. What operation will you use to convert? _____________

7. By what number do you need to multiply or divide? _____________

8. What is the answer to the problem? _____________

Holt Mathematics

LESSON 1-3 — Puzzles, Twisters, & Teasers
Measure for Measure

Choose the most appropriate metric unit for each measurement. Circle the letter above your answer.

1. The capacity of a juice can

Y	T
milliliters	kiloliters

2. the length of a swimming pool

H	O
millimeters	meters

3. The mass of a bowling ball

E	U
milligrams	kilograms

4. The capacity of a walk-in freezer

R	L
kiloliters	milliliters

5. The width of a postcard

E	I
centimeters	kilometers

6. The mass of a button

T	A
kilograms	grams

Choose the measurement that is equivalent to the one given. Circle the letter above your answer.

7. 680 mg = ?

T	R
0.068 g	0.68 g

8. 13.7 cm = ?

D	L
137 mm	13.7 m

9. 0.49 kg = ?

E	R
4,900 g	490 g

10. 27.1 L = ?

U	T
0.0271 kL	271 mL

11. 964 mm = ?

M	O
0.964 m	9.64 cm

12. 0.78 m = ?

T	S
78 mm	78 cm

Start with problem number 1, and use the circled letters to solve the riddle.

What part of your body has the most rhythm?

____ ____ ____ ____

____ ____ ____ ____ ____ __

Holt Mathematics

Practice A

Powers of Ten and Scientific Notation

Choose the letter for the best answer.

1. $4 \cdot 10^2$

  A 4   **C** 400

  B 40   **D** 100

2. $16 \cdot 10^0$

  F 1600   **H** 10

  G 16   **J** 0

3. $9 \cdot 10^3$

  A 9,000   **C** 90

  B 900   **D** 9

4. $17 \cdot 10^1$

  F 1.7   **H** 170

  G 17   **J** 1700

Multiply.

5. $23 \cdot 10^2$ **6.** $15 \cdot 10^4$ **7.** $30 \cdot 10^2$ **8.** $28 \cdot 10^3$

9. $132 \cdot 10^2$ **10.** $201 \cdot 10^3$ **11.** $456 \cdot 10^2$ **12.** $108 \cdot 10^4$

Write the number in scientific notation.

13. 56,000   **14.** 306,000   **15.** 8,000,000   **16.** 7,200,000

17. 14,000,000   **18.** 41.00   **19.** 2,144,000   **20.** $20.3 \cdot 10^5$

21. Lake Huron covers an area of about 23,000 square miles. Write this number in scientific notation.

22. The planet Mercury is about 3.6×10^7 miles from the sun. Write this number in standard form.

Holt Mathematics

Practice B
Powers of Ten and Scientific Notation

Multiply.

1. $6 \cdot 10^3$ **2.** $22 \cdot 10^1$ **3.** $8 \cdot 10^2$ **4.** $18 \cdot 10^0$

5. $70 \cdot 10^2$ **6.** $25 \cdot 10^3$ **7.** $3 \cdot 10^4$ **8.** $180 \cdot 10^3$

9. $84 \cdot 10^4$ **10.** $315 \cdot 10^2$ **11.** $210 \cdot 10^3$ **12.** $1{,}004 \cdot 10^3$

13. $1{,}764 \cdot 10^1$ **14.** $856 \cdot 10^0$ **15.** $4{,}055 \cdot 10^3$ **16.** $716 \cdot 10^4$

Write each number in scientific notation.

17. 34,000   **18.** 7,700   **19.** 2,100,000   **20.** 404,000

21. 21,000,000   **22.** 612.00   **23.** 3,001,000   **24.** $62.13 \cdot 10^4$

25. Lake Superior covers an area of about 31,700 square miles. Write this number in scientific notation.

26. Mars is about $1.42 \cdot 10^8$ miles from the sun. Write this number in standard form.

27. In 2005, the population of China was about $1.306 \cdot 10^9$. What was the population of China written in standard form?

28. A scientist estimates there are 4,800,000 bacteria in a test tube. How does she record the number using scientific notation?

Holt Mathematics

Practice C
Powers of Ten and Scientific Notation

Multiply.

1. $5 \cdot 10^3$

2. $471 \cdot 10^2$

3. $39.5 \cdot 10^1$

4. $200 \cdot 10^5$

5. $7{,}025 \cdot 10^0$

6. $5.7 \cdot 10^6$

7. $66.25 \cdot 10^4$

8. $9.01 \cdot 10^9$

Write each number in scientific notation.

9. 25,000

10. 9,900

11. 9,700,000

12. $95.6 \cdot 10^8$

13. 23,000,000,000

14. 110.00

15. $301.9 \cdot 10^5$

16. $73.55 \cdot 10^4$

Write the missing number or numbers.

17. $1.23 \times 10^? = 12{,}300$

18. $8.3 \times 10^5 = ?$

19. $112{,}000{,}000 = ? \times 10^8$

20. $410{,}000 = ? \times 10^5$

21. $7.7 \times 10^7 = ?$

22. $2{,}950{,}000 = 2.95 \times 10^?$

23. The Caspian Sea covers an area of about 143,250 square miles. Write this number in scientific notation.

24. The distance between Jupiter and Saturn is about 4.03×10^8 miles. Write this number in standard form.

25. In 2002, there were about 3.005×10^7 pet dogs in Brazil and about 9.65×10^6 pet dogs in Japan. In which country were there more pet dogs?

26. A scientist estimates there are 37,000,000 bacteria in a petri dish. He records the number using scientific notation. What does he write?

Holt Mathematics

LESSON 1-4 Reteach
Powers of Ten and Scientific Notation

To multiply by a power of 10, use the exponent to find the number of zeros in the product.

- Multiply $42 \cdot 10^4$.

$$42 \cdot 10^4 = 420,000$$

> The exponent **4** tells you to write **4** zeros after 42 in the product.

Find each product.

1. $84 \cdot 10^3$

The product should have _____ zeros.

$84 \cdot 10^3 =$ _____________

2. $61 \cdot 10^5$

The product should have _____ zeros.

$61 \cdot 10^5 =$ _____________

3. $22 \cdot 10^6$ **4.** $753 \cdot 10^3$ **5.** $825 \cdot 10^2$ **6.** $123 \cdot 10^1$

___________   ___________   ___________   ___________

- Write 926,000 in scientific notation.

First, write the digits before the zeros as a number greater than or equal to 1 and less than 10. The number must have only 1 digit to the left of the decimal point. That digit cannot be zero.

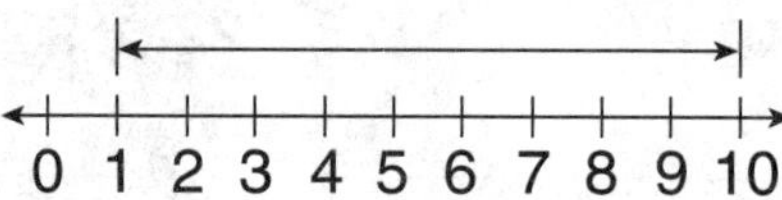

Think: 9.26 is greater than 1 and less than 10.

Then multiply 9.26 by the power of 10 that gives 926,000 as the product.

$$9\,2\,6\,0\,0\,0 = 9.26 \times 10^5$$

> The decimal point moves **5** places so the exponent is **5**.

Write each number in scientific notation.

7. 5,100

The decimal point moves _____ places.

$5,100 =$ _____ . _____ $\times\ 10^-$

8. 1,840,000

The decimal point moves _____ places.

$1,840,000 =$ _____ . _____ $\times\ 10^-$

9. 641,000   **10.** 47,300   **11.** 8,250,000   **12.** 703,000

___________   ___________   ___________   ___________

Holt Mathematics

<table><tr><td>LESSON
1-4</td><td>

Challenge
Computer Bytes
</td></tr></table>

Each byte in a computer's memory represents about one character.
The major units of computer memory are kilobytes (KB), megabytes
(MB), and gigabytes (GB).

1 kilobyte = 1,000 bytes		1 KB = 1,000 bytes
1 megabyte = 1,000 kilobytes		1 MB = 1,000 KB
1 gigabyte = 1,000 megabytes		1 GB = 1,000 MB

Write your answers using scientific notation.

1. In 1984, many personal computers
had 64 KB of active (RAM) memory.
How many bytes does this represent?

2. In 1992, many personal computers
had 40 MB of hard drive memory.
How many bytes does this represent?

3. In 1997, many personal computers
had 1 GB of hard drive memory. How
many bytes does this represent?

4. By 2005, many personal computers
had 250 GB of hard drive memory.
How many bytes does this represent?

Ming saved his computer files on floppy disks. Each disk holds up
to 1.44 MB of memory. He used these disks to transfer his files to
another computer.

5. How many bytes could each floppy
disk hold?

6. Ming's new computer has 120 GB of
memory. How many disks could he
transfer if each disk held 1.2 MB?

Rachel decided to back up her hard drive's computer files by
copying them onto compact disks (CDs). Each CD can hold up
to 650 MB of memory, but Rachel saves only 600 MB on each.

7. How many bytes could each CD
potentially hold?

8. If Rachel backs up 6 GB of memory,
how many bytes of memory will she
need?

Holt Mathematics

<table><tr><td>LESSON
1-4</td><td># Problem Solving
Powers of Ten and Scientific Notation</td></tr></table>

Write the correct answer.

1. Earth is about 150,000,000 kilometers from the sun. Write this distance in scientific notation.

2. The planet Neptune is about 4.5×10^9 kilometers from the sun. Write this distance in standard form.

3. At the end of 2004, the U.S. federal debt was about $7 trillion, 600 billion. Write the amount of the debt in standard form and in scientific notation.

4. Canada is about 1.0×10^7 square kilometers in size. Brazil is about 8,500,000 square kilometers in size. Which country has a greater area?

Choose the letter for the best answer.

5. China's population in 2001 was approximately 1,273,000,000. Mexico's population for the same year was about 1.02×10^8. How much greater was China's population than Mexico's?

A 1,375,000,000

B 1,274,020,000

C 1,171,000,000

D 102,000,000

6. In mid-2001, the world population was approximately 6.137×10^9. By 2050, the population is projected to be 9.036×10^9. By how much will world population increase?

F 151,730,000

G 289,900,000

H 1,517,300,000

J 2,899,000,000

7. The Alpha Centauri star system is about 4.3 light-years from Earth. One light-year, the distance light travels in 1 year, is about 6 trillion miles. About how many miles away from Earth is Alpha Centauri?

A 2.58×10^{13} miles

B 6×10^{13} miles

C 1.03×10^{12} miles

D 2.58×10^9 miles

8. In the fall of 2001, students in Columbia, South Carolina, raised $440,000 to buy a new fire truck for New York City. If the money had been collected in pennies, how many pennies would that have been?

F 4.4×10^6

G 4.4×10^5

H 4.4×10^7

J $4.4 \; 3 \; 10^8$

Holt Mathematics

<table><tr><td>LESSON
1-4</td><td>

Reading Strategies
Follow a Procedure
</td></tr></table>

When you have an exponent with a base of 10, the number is called
a **power of 10.**
You can use some simple rules to find products with powers of 10.

13×10^3 32.5×10^3

↓   ↓

$13 \times (10 \times 10 \times 10)$ $32.5 \times (10 \times 10 \times 10)$

↓   ↓

$13 \times 1{,}000$ $32.5 \times 1{,}000$

↓   ↓

3.0.0.0.   32,500.

Move the decimal point   Move the decimal point
3 places to the right.   3 places to the right.
You need to add 3 zeros.   You need to add 2 zeros.

You can use powers of 10 to write large numbers in **scientific
notation.** Scientific notation is used as a shortcut to write very large
or very small numbers.

To write 268,000,000 in scientific notation:

Step 1: Move the decimal point to create   2.6.8.0.0.0.0.0.(⟵ Move the decimal
a number between 1 and 10.   point 8 places.

Step 2: The number of places the 2.68×10^8 ⟵ So, the exponent
decimal point is moved is the   is 8.
value of the exponent.

268,000,000 written in scientific notation is 2.68×10^8.

Use 2.8×10^5 to answer Exercises 1–4.

1. How many times is 10 a factor? _________________________________

2. Rewrite the number with 10 as a repeated factor.

____

3. How many places will you move the decimal point?
How many zeros will be in the product?   _________ _________

4. What is the product of 2.8×10^5?   _____________________

Holt Mathematics

<table><tr><td>LESSON
1-4</td><td>

Puzzles, Twisters & Teasers
Oh, the Power of Tens!
</td></tr></table>

Substitute the correct number for the letter or letters in each equation. Use your answers to solve the riddle.

1. $24{,}500 = 2.45 \times 10^E$ _____________

2. $280{,}000 = 2.8 \times 10^P$ _____________

3. $592{,}000 = I \times 10^5$ _____________

4. $16{,}800 = C \times 10^4$ _____________

5. $5.4 \times 10^H = 540{,}000{,}000$ _____________

6. $24{,}400{,}000 = S \times 10^A$ _____________

What's a Martian's favorite snack?

___ ___ ___ ___ ___   ___ ___ ___ ___ ___
2.44   5   7   1.68   4   1.68   8   5.92   5   2.44

34

Holt Mathematics

Practice A

LESSON 1-5

Order of Operations

Choose the letter for the best answer.

1. $75 + 12 \cdot 2$

   A 87   **C** 108

   B 99   **D** 174

2. $100 - 25 \div 5$

   F 15   **H** 80

   G 75   **J** 95

3. $50 - 18 \div 6 + 2$

   A 49   **C** 10

   B 40   **D** 4

4. $72 - 4^2 \cdot 2$

   F 32   **H** 56

   G 40   **J** 64

5. $(8 + 22) \div 5 + 5$

   A 30   **C** 11

   B 17.4   **D** 3

6. $3^3 - (9 \cdot 2 + 1)$

   F 19   **H** 8

   G 10   **J** −10

Simplify each expression.

7. $2^4 \div 8 + 5$

8. $18 + 2(1 + 3^2)$

9. $(16 \div 4) + 4 \cdot (2^2 - 2)$

_________   _________   _________

10. $2^3 - (3 \cdot 5 - 8)$

11. $35 + 4^2 - (6 - 3)$

12. $6 \cdot 7 - 3(4 + 1)$

_________   _________   _________

13. $100 \div 5 \cdot 2^2$

14. $(5 + 2)^2 \div 7 - 6$

15. $15 - 3 \cdot 4 \div 2 + 5$

_________   _________   _________

16. Leon rents a video game for $5. He returns the video game 3 days late. The late fee is $1 for each day late. Simplify the expression $5 + 3 \cdot 1$ to find out how much it costs Leon to rent the video game.

____

17. Olivia shovels snow for 6 neighbors. Each neighbor pays her $8. Two neighbors each give her a $2 tip as well. Simplify the expression $6 \cdot 8 + 2 \cdot 2$ to find out how much money Olivia earns in all.

____

Holt Mathematics

Practice B
Order of Operations

Simplify each expression.

1. $15 \cdot 3 + 12 \cdot 2$

2. $212 + 21 \div 3$

3. $9 \cdot 3 - 18 \div 3$

____

4. $65 - 36 \div 3$

5. $100 - 9^2 + 2$

6. $3 \cdot 5 - 45 \div 3^2$

____

7. $54 \div 6 + 4 \cdot 6$

8. $(6 + 5) \cdot 16 \div 2$

9. $60 - 8 \cdot 12 \div 3$

____

10. $45 - 3^2 \cdot 5$

11. $52 - (8 \cdot 2 \div 4) + 3^2$

12. $(2^3 + 10 \div 2) \cdot 3$

____

13. $25 + 7(18 - 4^2)$

14. $(6 \cdot 3 - 12)^2 \div 9 + 7$

15. $4^3 - (3 + 12 \cdot 2 - 9)$

____

16. $2^4 \div 8 + 5$

17. $(1 + 2)^2 \cdot (3 - 1)^2 \div 2$

18. $(16 \div 4) + 4 \cdot (2^2 - 2)$

____

19. $2^5 - (3 \cdot 7 - 7)$

20. $75 + 5^2 - (8 - 3)$

21. $9 \cdot 6 - 5(10 - 3)$

____

22. $96 \div 4 + 5 \cdot 2^2$

23. $(15 - 6)^2 \div 3 - 3^3$

24. $19 - 8 \cdot 5 \div 10 + 6 \div 3$

____

25. Jared has $32. He buys 5 packs of trading cards that cost $3 each and a display book that costs $7. Simplify the expression $32 - (5 \cdot 3 + 7)$ to find out how much money Jared has left.

____

26. David buys 3 movie tickets for $6 each and 2 bags of popcorn for $2 each. Simplify the expression $3 \cdot 6 + 2 \cdot 2$ to find out how much money David spent in all.

____

Holt Mathematics

Practice C
LESSON 1-5 *Order of Operations*

Simplify each expression.

1. $25 \cdot 3 + 60 \cdot 2$

2. $350 \div 5 + 12 \cdot 7$

3. $3 \cdot 9 + 96 \div 4$

4. $77 - 42 \div 7^1$

5. $532 - 2^5 \div 4$

6. $3(20 - 4^2) + 7$

7. $270 \div 6 + 6^2$

8. $(5 + 6)^2 + 18 \div 2$

9. $10^2 - 25 \cdot 3 \div 5$

10. $65 - 4^3 \cdot 1^7$

11. $40 - (5 \cdot 2) + 8$

12. $(6^2 + 4) \div 5$

13. $2^4 \div 8 + 5$

14. $(1 + 2)^2 \cdot (3 - 1)^2 \div 2$

15. $(16 \div 4) + 4 \cdot (2^2 - 2)$

Insert grouping symbols to make each statement true.

16. $18 + 2 \cdot 1 + 3^2 = 38$

17. $4 \cdot 2 - 2^2 \div 9 + 2 = 6$

18. $3^3 - 9 \cdot 2 + 1 = 8$

19. $2^3 - 3 \cdot 5 - 8 = 1$

20. $35 + 4^2 - 6 - 3 = 48$

21. $6 \cdot 7 - 3 \cdot 4 + 1 = 27$

22. A group of students charges \$7 to clean the exterior and \$6 to clean the interior of a car. They clean 9 exteriors and 5 interiors. Simplify the expression $7 \cdot 9 + 6 \cdot 5$ to find out how much money the students raised in all.

23. Ariel has \$65. She buys 5 books that cost \$8.00 each, a bookmark that costs \$2.00, and a magazine that costs \$4.00. Simplify the expression $65 - (5 \cdot 8 + 2 + 4)$ to find out how much money Ariel has left.

Holt Mathematics

To help you remember the order of operations use the phrase
"**P**lease **E**xcuse **M**y **D**ear **A**unt **S**ally."

> **P**: first, **p**arentheses (if any)
> **E**: second, **e**xponents (if any)
> **M** and **D**: then, **m**ultiplication and **d**ivision, in order from left to right
> **A** and **S**: finally, **a**ddition and **s**ubtraction, in order from left to right

Evaluate.

$$39 \div (9 + 4) + 5 - 2^2$$

Parentheses $\longrightarrow$ $39 \div 13 + 5 - 2^2$

Exponents $\longrightarrow$ $39 \div 13 + 5 - 4$

Multiply and divide from left to right $\longrightarrow$ $3 + 5 - 4$

Add and subtract from left to right $\longrightarrow$ $8 - 4 = 4$

Simplify each expression.

1. $12 \cdot 4 - 2$

_____ $- 2$

_____

2. $15 \div 3 \cdot 5$

_____ $\cdot 5$

_____

3. $15 \cdot 3 \div 5$

_____ $\div 5$

_____

4. $8 + 20 \div 4$

_____

5. $5 - 2 \cdot 6 \div 4 + 1$

_____

6. $3^2 + 6 \cdot 4 - 5^2$

_____

7. $1 + 4 \cdot 9 \div 6 - 7$

_____

8. $18 \div (6 \div 3)$

_____

9. $(18 \div 6) \div 3$

_____

10. $4 \cdot 5 + 8 \div 2 - 7$

_____

11. $2 \cdot 3 - 8 \div 2^2$

_____

12. $8(7 - 6) \div 2^3$

_____

Holt Mathematics

LESSON 1-5 Challenge
Fixed and Variable Costs

A *fixed cost* is a one-time cost. A *variable cost* changes depending on your use of a product or a service.

The annual enrollment fee per year at a fitness club (fixed cost) is $30. You also pay $2 per visit (variable cost), and you visit the club 8 times per month. What is your total annual cost?

$$2 \cdot (8 \cdot 12)$$ variable cost times total visits

$$30 + 2 \cdot (8 \cdot 12)$$ total annual cost

$$222$$ Your total annual cost is $222.

Use the information above to solve problems 1–4.

1. Suppose the annual fee is $25, but the cost per visit is $3. What is your annual cost?

2. Suppose you visit the club 3 times per week instead of 8 times per month. What is your annual cost?

3. Suppose the cost per visit is $3 after the first 50 visits per year. What is your annual cost?

4. Suppose you pay for up to 75 visits per year. Any additional visits are free. What is your annual cost?

The school band is raising money for a trip. The members ordered 5 dozen jerseys for $7 each and sold them for $12 each. They also ordered 4 dozen sweatshirts for $11 each and sold them for $18 each. The band paid $35 to create the design.

5. Write and simplify an expression to calculate the band's variable costs for the clothing.

6. Write and simplify an expression to calculate the band's total cost, including fixed costs.

7. Write and simplify an expression to calculate the band's profit.

8. What would the profit be if jerseys sold for $10 and sweatshirts for $20?

Holt Mathematics

Name _________________________________ Date _________ Class ___________

Problem Solving
Order of Operations

Write the correct answer.

1. In 1975, the minimum wage was $2.10 per hour. Write and simplify an expression to show wages earned in a 35-hour week after a $12 tax deduction.

2. George bought 3 boxes of Girl Scout cookies at $3.50 per box and 4 boxes at $3.00 per box. Write and simplify an expression to show his total cost.

3. In 1 week Ed works 4 days, 3 hours a day, for $12 per hour, and 2 days, 6 hours a day, for $15 per hour. Simplify the expression $12(4 \cdot 3) + 15(2 \cdot 6)$ to find Ed's weekly earnings.

4. Keisha had $150. She bought jeans for $27, a sweater for $32, 3 blouses for $16 each, and 2 pairs of socks for $6 each. Simplify the expression $150 - [27 + 32 + (3 \cdot 16) + (2 \cdot 6)]$ to find out how much money she has left.

Choose the letter for the best answer.

5. As of September 1, 1997, the minimum wage was set at $5.15 per hour. How much more would someone earn now than in 1997 if she earns $5 more per hour for a 40-hour week?

   A $206 more

   B $200 more

   C $406 more

   D $400 more

6. Gary received $200 in birthday gifts. He bought 5 CDs for $15 each, 2 posters for $12 each, and a $70 jacket. How much money does he have left?

   F $31

   G $10

   H $132

   J $169

7. Yvonne took her younger brother and his friends to the movies. She bought 5 tickets for $8 each, 4 drinks for $2 each, and two $3 containers of popcorn. How much did she spend?

   A $22

   B $51

   C $54

   D $38

8. On a business trip, Mr. Chang stayed in a hotel for 7 nights. He paid $149 per night. While he was there, he made 8 phone calls at $2 each and charged $81 to room service. How much did he spend?

   F $246

   G $946

   H $1,043

   J $1,140

Holt Mathematics

LESSON 1-5 Reading Strategies
Use a Flowchart

When you read a book, you read from left to right. When you
evaluate an expression, you cannot always work from left to right.
You must follow a special rule called the **order of operations.** Use
the flowchart below to help you follow the order of operations.

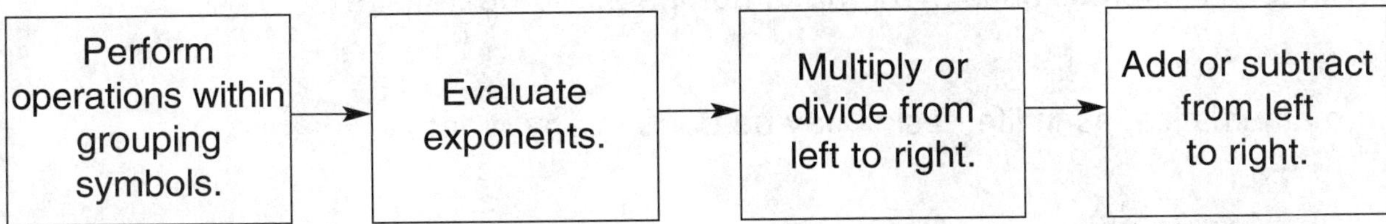

Example $25 - (4 \cdot 5) \div 2^2$ parentheses first

$25 - 20 \div 2^2$ Evaluate power next.

$25 - 20 \div 4$ Multiply and divide (left to right).

$25 - 5$ Add and subtract (left to right).

20

Answer each question.

1. In what order will you perform the operations in the following
   expression: $30 - 18 \div 2 \cdot 3 + 4$?

   ____

2. Simplify this expression: $30 - 18 \div 2 \cdot 3 + 4.$ ___________________

3. Make a flow chart for the order of operations in this expression:
   $3^3 - (2 + 5 \cdot 2).$

   ____

4. Simplify this expression: $3^3 - (2 + 5 \cdot 2).$ ___________________

5. Make a flow chart for the order of operations in this expression:
   $(3 + 12 \div 3)^2 - 4.$

   ____

6. Simplify this expression: $(3 + 12 \div 3)^2 - 4.$ ___________________

Holt Mathematics

Name ____ Date __________ Class __________

Puzzles, Twisters & Teasers
Is Everything in Order?

What's the last thing you take off before you go to bed?

Decide whether each statement below is true or false. Use your answers to solve the riddle.

1. A numerical expression is made up of numbers and operations.

   T   **F**

2. In mathematics, as in life, tasks may be done in any order.

   T   **F**

3. When using the order of operations, you should do division after subtraction.

   T   **F**

4. When using the order of operations, you should subtract and add from left to right.

   T   **F**

5. When using the order of operations, you should divide and multiply from right to left.

   T   **F**

6. When an expression has a set of grouping symbols within a second set of grouping symbols, you should begin with the innermost set.

   T   **F**

7. You should perform operations inside parentheses first.

   T   **F**

8. When using the order of operations, you should evaluate the power expression after multiplying and adding.

   T   **F**

9. Mathematicians agree on using the order of operations.

   T   **F**

T
L F Y
R H

F
O T
E U

You take ____ ____ ____ ____ ____ ____ ____ ____ ____ ____ ____ ____
         T   F   F   T   T   F   F   F   F   T   T

____ ____ ____ ____ ____ ____ ____ ____ ____ ____.
 F   T   F   T   T   F   F   T

42

Holt Mathematics

Practice A
Properties

Tell which property is shown.

1. $5 + 0 = 1$

____

2. $8 \bullet (6 \bullet 2) = (8 \bullet 6) \bullet 2$

____

3. $9 + 8 = 8 + 9$

____

4. $4 \bullet 1 = 4$

____

Simplify each expression. Write a reason for each step.

5. $13 + 28 + 7$

$13 + 28 + 7 = 28 + 13 + 7$ Reason: Commutative Property

$= 28 + (13 + 7)$ Reason: _______________________

$= 28 +$ _______   Reason: Add.

$=$ _______   Reason: _______________________

6. $20 \bullet (17 \bullet 5)$

$20 \bullet (17 \bullet 5) = 20 \bullet ($_______ $\bullet\ 17)$ Reason: _______________________

$= (20 \bullet$ _______ $) \bullet 17$ Reason: _______________________

$=$ _______ $\bullet$ _______   Reason: Multiply.

$=$ _______   Reason: _______________________

Use the Distributive Property to find each product.

7. $4(17)$

$4(17) = 4 \bullet (10 +$ _______ $)$

$= (4 \bullet$ _______ $) + (4 \bullet 7)$

$=$ _______ $+$ _______

$=$ _______

8. $3(28)$

$3(28) =$ _______________________

$=$ _______________________

$=$ _______________________

$=$ _______________________

Holt Mathematics

LESSON 1-6 — Practice B
Properties

Tell which property is represented.

1. $12 \cdot 14 = 14 \cdot 12$

2. $1 \cdot 36 = 36$

3. $(17 + 36) + 4 = 17 + (36 + 4)$

4. $8 \cdot 12 \cdot 5 = 8 \cdot (12 \cdot 5)$

Simplify each expression. Justify each step.

5. $4 \cdot 9 \cdot 50$

$4 \cdot 9 \cdot 50 = $ _________________________

$= $ _________________________

$= $ _________________________

$= $ _________________________

6. $(33 + 45) + 7$

$(33 + 45) + 7 = $ _________________________

$= $ _________________________

$= $ _________________________

$= $ _________________________

Use the Distributive Property to find each product.

7. $3(26) = $ _________________________

8. $(18)9 = $ _________________________

$= $ _________________________ $= $ _________________________

$= $ _________________________ $= $ _________________________

$= $ _________________________ $= $ _________________________

Holt Mathematics

Practice C
Properties

Complete each equation. Then tell which property is represented.

1. $23 +$ _______ $= 23$

____

2. _______ $\cdot (19 + 6) = (7 \cdot 19) + (7 \cdot 6)$

____

3. $27 + 45 =$ _______ $+ 27$

____

4. $6 \cdot ($_______ $\cdot 7) = (6 \cdot 14) \cdot 7$

____

Simplify each expression. Justify each step.

5. $(40 \cdot 7) \cdot 5$

$(40 \cdot 7) \cdot 5 =$ _______________________________

$=$ _______________________________

$=$ _______________________________

$=$ _______________________________

6. $15 + 98 + 85$

$15 + 98 + 85 =$ _______________________________

$=$ _______________________________

$=$ _______________________________

$=$ _______________________________

Use the Distributive Property to find each product.

7. $7(43) =$ _______________

$=$ _______________

$=$ _______________

$=$ _______________

8. $(597)4 =$ _______________

$=$ _______________

$=$ _______________

$=$ _______________

Holt Mathematics

LESSON 1-6 **Reteach**
Properties

You can use the Commutative Property, the Associative Property, and the Distributive Property with mental math to simplify expressions.

$16 + 47 + 14 = 47 + 16 + 14$	Commutative Property	$8 \cdot 3 \cdot 5 = 3 \cdot 8 \cdot 5$
$= 47 + (16 + 14)$	Associative Property	$= 3 \cdot (8 \cdot 5)$
$= 47 + 40$	Mental math	$= 3 \cdot 40$
$= 87$	Mental math	$= 120$

$9(28) = 9(20 + 8)$		$9(28) = 9(30 - 2)$
$= (9 \cdot 20) + (9 \cdot 8)$	Distributive Property	$= (9 \cdot 30) - (9 \cdot 2)$
$= 180 + 72$	Mental math	$= 270 - 18$
$= 252$	Mental math	$= 252$

Simplify each expression. Tell what properties you used.

1. $(45 + 39) + 25 = (39 + \underline{\quad}) + 25$ _________________ Property

$= 39 + (\underline{\quad} + \underline{\quad})$ _________________ Property

$= 39 + \underline{\quad}$

$= \underline{\quad}$

2. $25 \cdot 7 \cdot 4 = 25 \cdot \underline{\quad} \cdot \underline{\quad}$ _________________ Property

$= (\underline{\quad} \cdot \underline{\quad}) \cdot \underline{\quad}$ _________________ Property

$= \underline{\quad} \cdot \underline{\quad}$

$= \underline{\quad}$

3. 3. $5(18) = 5 \cdot (10 + \underline{\quad})$ **4.** $6(29) = 6 \cdot (30 - \underline{\quad})$

$= (5 \cdot \underline{\quad}) + (5 \cdot \underline{\quad})$ $= (6 \cdot \underline{\quad}) - (6 \cdot \underline{\quad})$

$= \underline{\quad} + \underline{\quad}$ $= \underline{\quad} - \underline{\quad}$

$= \underline{\quad}$ $= \underline{\quad}$

_________________ Property   _________________ Property

Holt Mathematics

Challenge
What's the Expression?

On each tic-tac-toe board below, exactly one row, column, or
diagonal contains three expressions with the same value.

**Predict which row, column, or diagonal of squares will have
three expressions with the same value. Draw a line through
those three squares. Then check your prediction. Find the
value of each expression on the board.**

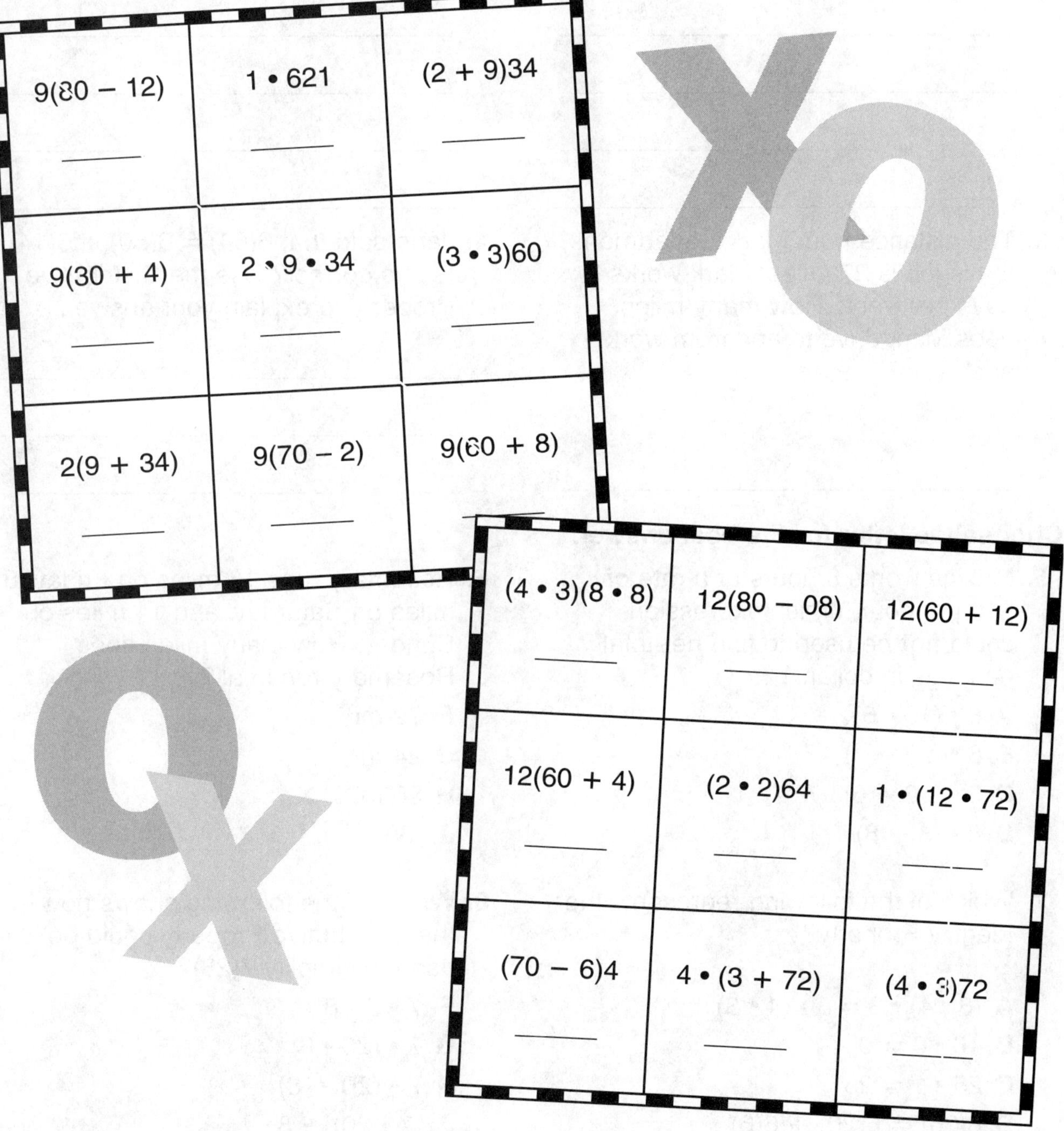

Holt Mathematics

Name _______________________________________ Date ___________ Class ___________

Problem Solving
Properties

Write the correct answer.

1. Jo makes and sells jewelry. She sold three bracelets for $45, $17, and $25. Write an expression for the total Jo received. Explain how you can use properties and mental math to simplify the expression.

2. Use parentheses to show two ways of grouping the numbers in 12 • 8 • 25. Tell which expression you think would be easier to simplify, and why. Then simplify the expression.

3. The distance from Mark's apartment to his job is 27 miles. Mark works 5 days per week. How many miles does Mark drive to and from work each week?

4. Jane said that 6(64) = 6(50) + 6(14). Is she correct? Use the Distributive Property to explain your answer.

Choose the letter for the best answer.

5. Maxine works 8 hours at a rate of $16 per hour. Which expression could **not** be used to find her total earnings in dollars?

   A 8 • (10 • 6)

   B 8 • (20 − 4)

   C 8 • (10 + 6)

   D 8 • (8 + 8)

6. Rosemary runs 16 miles on Friday, 8 miles on Saturday, and 14 miles on Sunday. How many miles does Rosemary run in all?

   F 22 mi

   G 24 mi

   H 30 mi

   J 38 mi

7. Which of the following represents the Identity Property?

   A (8 • 4) • 3 = 8 • (4 • 3)

   B 16 • 0 = 0

   C 25 • 1 = 25

   D 6(26) = 6(20) + 6(6)

8. Which of the following shows how the Distributive Property could be used to simplify 7(28)?

   F 7 • 2 • 8

   G 7 • (20 • 8)

   H 7 • (20 + 8)

   J (7 • 20) + 8

Holt Mathematics

Reading Strategies
Use a Flowchart

Use a flowchart to help you simplify an expression, such as $(25 + 89) + 15$.

Step 1: Choose two numbers that are easy to add.
$(25 + 89) + 15$

Step 2: Rewrite the expression so the two numbers are next to each other.
Use the Commutative Property. $(25 + 89) + 15 = (89 + 25) + 15)$

Step 3: Rewrite the expression so the two numbers are grouped together.
Use the Associative Property. $(89 + 25) + 15 = 89 + (25 + 15)$

Step 4: Add.
$89 + (25 + 15) = 89 + 40 = 129$

Use the expression $16 + (39 + 14)$ for Exercises 1-4.

1. Which two numbers are easy to add? _______________________

2. Rewrite the expression so that the numbers that are easy
   to add are next to each other. What property lets you do this?

   ____

3. Rewrite the expression so that the numbers that are easy to add
   are grouped together. What property lets you do this?

   ____

4. Simplify the expression.

   ____

Use the expression $35 + 47 + 5$ for Exercises 5-8.

5. Which two numbers are easy to add?

   ____

6. Rewrite the expression so that the numbers that are easy to add
   are next to each other. What property lets you do this?

   ____

7. Rewrite the expression so that the numbers that are easy to add
   are grouped together. What property lets you do this?

   ____

8. Simplify the expression.

   ____

Holt Mathematics

<table><tr><td>**LESSON**
1-6</td><td></td></tr></table>

Puzzles, Twisters, & Teasers
Make the Connection!

Draw a line to connect each equation to the property it represents.

1. $3 \cdot (8 \cdot 5) = (3 \cdot 8) \cdot 5$ Identity Property   **R**

2. $1 \cdot 15 = 15$ Commutative Property   **Y**

3. $9(4 + 6) = 9(4) + 9(6)$ Associative Property   **E**

4. $38 + 7 = 7 + 38$ Distributive Property   **P**

Draw line to connect expressions with the same value.

5. $5 + (7 + 6)$ $5(7) + 5(6)$ **S**

6. $5(70 + 6)$ $5(70) - 5(6)$ **H**

7. $5 + 70 + 6$ $(5 + 7) + 6$ **T**

8. $5 \cdot 7 \cdot 6$ $(5 \cdot 70) \cdot 6$ **I**

9. $5(13)$ $7 \cdot (5 \cdot 6)$ **D**

10. $5(70 - 6)$ $70 + 11$ **L**

11. $5 \cdot (70 \cdot 6)$ $5(70) + 5(6)$ **K**

Start with number 1. Find the letter next to the answer. Use the letters to find out why potatoes are good detectives.

___ ___ ___ ___   ___ ___ ___ ___
 5   10   1   4   6   1   1   3

___ ___ ___ ___ ___   ___ ___ ___ ___
 5   10   1   11   2   1   4   1   9

___ ___ ___ ___ ___ ___
 3   1   1   7   1   8

Holt Mathematics

Practice A
Variables and Algebraic Expressions

Find the value of $n + 3$ for each value of n.

1. $n = 4$ **2.** $n = 7$ **3.** $n = 0$ **4.** $n = 32$

_________   _________   _________   _________

Find the value of $x - 9$ for each value of x.

5. $x = 12$ **6.** $x = 57$ **7.** $x = 19$ **8.** $x = 100$

_________   _________   _________   _________

Find the value of each expression using the given value for each variable.

9. $3n$ for $n = 4$ **10.** $x + 8$ for $x = 8$ **11.** $9p - 6$ for $p = 2$

_________   _________   _________

12. $n \div 5$ for $n = 35$ **13.** $6x + 18$ for $x = 0$ **14.** $s - 7$ for $s = 8$

_________   _________   _________

15. $3w + 5$ for $w = 3$ **16.** $c - 9$ for $c = 12$ **17.** $2a \div 3$ for $a = 6$

_________   _________   _________

18. $y + z$ for $y = 10$ and $z = 20$ **19.** $3w - 2v$ for $w = 7$ and $v = 8$

_________   _________

20. $4a \div b$ for $a = 6$ and $b = 4$ **21.** $5s + 4t$ for $s = 3$ and $t = 4$

_________   _________

22. The expression $7w$ gives the number of days in w weeks. Find the value of $7w$ for $w = 20$. How many days are there in 20 weeks? _________________

23. A cat can run as fast as $m \div 2$ miles per minute in m minutes. Find the value of $m \div 2$ for $m = 10$. How many miles can a cat run in 10 minutes? _________________

24. Tyrone works 8 hours a day. You can use the expression $8d$ to find the total number of hours he works in d days. How many hours does he work in 5 days? _________________

Holt Mathematics

Practice B
Variables and Algebraic Expressions

Evaluate $n - 5$ for each value of n.

1. $n = 8$ **2.** $n = 121$ **3.** $n = 32$ **4.** $n = 59$

Evaluate each expression for the given values of the variable.

5. $3n + 15$ for $n = 4$ **6.** $h \div 12$ for $h = 60$ **7.** $32x - 32$ for $x = 2$

8. $\frac{c}{2}$ for $c = 24$ **9.** $(n \div 2)5$ for $n = 14$ **10.** $8p + 148$ for $p = 15$

11. $e^2 - 7$ for $e = 8$ **12.** $3d^2 + d$ for $d = 5$ **13.** $40 - 4k^3$ for $k = 2$

14. $2y - z$ for $y = 21$ and $z = 19$ **15.** $3h^2 + 8m$ for $h = 3$ and $m = 2$

16. $18 \div a + b \div 9$ for $a = 6$ and $b = 45$ **17.** $10x - 4y$ for $x = 14$ and $y = 5$

18. You can find the area of a rectangle with the expression lw where l represents the length and w represents the width. What is the area of the rectangle at right in square feet?

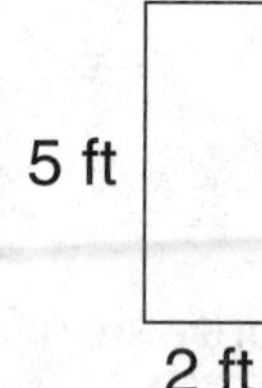

19. Rita drove an average of 55 mi/h on her trip to the mountains. You can use the expression $55h$ to find out how many miles she drove in h hours. If she drove for 5 hours, how many miles did she drive?

Holt Mathematics

Practice C
Variables and Algebraic Expressions

Evaluate each expression for the given values of the variable.

1. $3n + 4n$ for $n = 8$

2. $\dfrac{6s}{5}$ for $s = 25$

3. $q^2 + 5q - 11$ for $q = 4$

4. $\dfrac{350}{d} + 4d + 7$ for $d = 10$

5. $9x + 2x^2 + 2$ for $x = 2$

6. $8m^2 + 7 - 2m$ for $m = 3$

7. $4(h + k)$ for $h = 3$ and $k = 55$

8. $\dfrac{6r}{4} + 5s$ for $r = 8$ and $s = 18$

9. $6a - 2b^2$ for $a = 9$ and $b = 5$

10. $6h - 20g$ for $h = 1{,}500$ and $g = 200$

11. $\dfrac{36}{m^2} + \dfrac{n^2}{4}$ for $m = 6$ and $n = 16$

12. $x^2 - 2x - y^2$ for $x = 15$ and $y = 2$

13. $4d^3 + 6e^2 - \dfrac{8d}{2}$ for $d = 2$ and $e = 3$

14. $\dfrac{5r^2}{4} + \dfrac{4s}{3r}$ for $r = 4$ and $s = 9$

15. You can find the volume of a rectangular prism with the expression $a \bullet b \bullet c$, where a is the length, b is the width, and c is the height of the prism. What is the volume of the prism at the right in cubic inches?

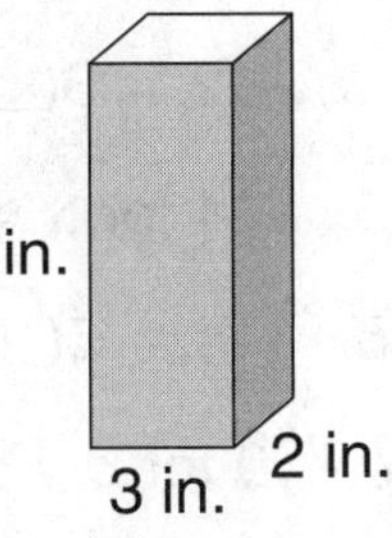

16. You can use the expression $5m$ to find out how many seconds it takes a sound to travel m miles through the air. However, through water, sound takes m seconds to travel m miles. Use the expression $5m - m$ to find out how much longer will it take a sound to travel 8 miles in air than in water.

Holt Mathematics

 Reteach

1-7 *Variables and Algebraic Expressions*

A **variable** is a letter that represents a number than can change in an expression. When you **evaluate** an algebraic expression, you substitute the value given for the variable in the expression.

- Algebraic expression: $x - 3$

  The value of the expression depends on the value of the variable x.

  If $x = 7 \rightarrow 7 - 3 = 4$

  If $x = 11 \rightarrow 11 - 3 = 8$

  If $x = 15 \rightarrow 15 - 3 = 12$

- Evaluate $4n + 1$ for $n = 5$.

  Replace the variable n with 5. $\rightarrow 4(5) + 1 = 20 + 1 = 21$

Evaluate each expression for the given value.

1. $a + 7$ for $a = 3$

  $a + 7 = 3 + 7 = $ ______

2. $k - 5$ for $k = 13$

  $k - 5 = $ ______ $- 5 = $ ______

3. $y \div 3$ for $y = 6$

  $y \div 3 = $ ______ $\div 3 = $ ______

4. $12 + m$ for $m = 9$

  $12 + m = $ ______ $+ $ ______ $= $ ______

5. $3n - 2$ for $n = 5$

  $3n - 2 = 3($______$) - 2 = $ ______ $- 2 = $ ______

6. $5x + 4$ for $x = 4$

  $5x + 4 = 5($______$) + $ ______ $= $ ______ $+ $ ______ $= $ ______

7. $c - 9$ for $c = 11$

8. $b + 16$ for $b = 4$

9. $a - 4$ for $a = 9$

______

______

______

10. $25 - g$ for $g = 12$

11. $w + 5$ for $w = 2$

12. $3 + s$ for $s = 8$

______

______

______

13. $7q$ for $q = 10$

14. $2y + 9$ for $y = 8$

15. $6x - 3$ for $x = 1$

______

______

______

Holt Mathematics

LESSON	**Challenge**
1-7	*What an Expression!*

**Complete each table with four expressions that have the same
value. Use each given value of *n*.**

1.

Expression Value 32	Value of *n* 4
Addition expression:	
Subtraction expression:	
Multiplication expression:	
Division expression:	

2.

Expression Value 96	Value of *n* 12
Addition expression:	
Subtraction expression:	
Multiplication expression:	
Division expression:	

3.

Expression Value 156	Value of *n* 6
Addition expression:	
Subtraction expression:	
Multiplication expression:	
Division expression:	

4.

Expression Value 98	Value of *n* 14
Addition expression:	
Subtraction expression:	
Multiplication expression:	
Division expression:	

5.

Expression Value 57	Value of *n* 12
Addition expression:	
Subtraction expression:	
Multiplication expression:	
Division expression:	

6.

Expression Value 248	Value of *n* 124
Addition expression:	
Subtraction expression:	
Multiplication expression:	
Division expression:	

Holt Mathematics

Problem Solving
Variables and Expressions

Write the correct answer.

1. In 2000, people in the United States watched television an average of 29 hours per week. Use the expression $29w$ for $w = 4$ to find out about how many hours per month this is.

2. Find the value of the variable w in the expression $29w$ to find the average number of hours people watched television in a year. Find the value of the expression.

3. The expression $y + 45$ gives the year when a person will be 45 years old, where y is the year of birth. When will a person born in 1992 be 45 years old?

4. The expression $24g$ gives the number of miles Guy's car can travel on g gallons of gas. If the car has 6 gallons of gas left, how much farther can he drive?

Choose the letter for the best answer.

5. Sam is 5 feet tall. The expression $0.5m + 60$ can be used to calculate his height in inches if he grows an average of 0.5 inch each month. How tall will Sam be in 6 months?

   A 56 inches

   B 5 feet 6 inches

   C 63 inches

   D 53 inches

6. The winner of the 1911 Indianapolis 500 auto race drove at a speed of about $s - 56$ mi/h, where s is the 2001 winning speed of about 131 mi/h. What was the approximate winning speed in 1911?

   F 75 mi/h

   G 186 mi/h

   H 85 mi/h

   J 187 mi/h

7. The expression $1{,}587v$ gives the number of pounds of waste produced per person in the United States in v years. How many pounds of waste per person is produced in the United States in 6 years?

   A 1,581 pounds

   B 1,593 pounds

   C 9,348 pounds

   D 9,522 pounds

8. The expression $\$1.25p + \$3.50$ can be used to calculate the total charge for faxing p pages at a business services store. How much would it cost to fax 8 pages?

   F $12.50

   G $4.75

   H $13.50

   J $10.00

Holt Mathematics

Reading Strategies

LESSON 1-7

Focus on Vocabulary

Your age varies from one year to the next. The cost of gasoline can vary from one week to the next. The word **vary** means change. In mathematical expressions, the value of a letter can change, or vary, so letters are called **variables.**

The opposite of vary is to stay **constant.** A constant value never changes.

An algebraic expression is made up of constants, variables, and operation symbols.

> algebraic expression = constants + variables + operation symbols
>
> Examples: $4x + 3y - 2^2$ $6p - 3t + 12$

To evaluate an algebraic expression, you need to:

Step 1: Substitute a value for each variable.   **Step 2:** Follow the order of operations.

Evaluate: $4a - 2b + 8$ for $a = 5$ and $b = 3$.

$4(5) - 2(3) + 8$ ←——————— Substitute 5 for a and 3 for b.
$20 - 6 + 8$ ←——————— Multiply first.
$14 + 8$ ←——————— Add and subtract from left to right.
22

Use this expression for Exercises 1–7: $8 + 4p - 2t$.

1. Write the variables in this expression. _______________ _______________

2. Rewrite the expression by substituting these values:
   $p = 6$ and $t = 2$.

   ____

3. What operation will you perform first? _______________________________

4. Perform this operation and rewrite the expression.

   ____

5. What operation will you perform next? _______________________________

6. Perform this operation and rewrite the expression. _______________________

Holt Mathematics

Puzzles, Twisters & Teasers

LESSON 1-7 *Movie Math!*

Circle words from the list in the word search. Then find an extra word in the word search that best completes the riddle.

expression   substitute   variable   value

constant   evaluate   algebra   algebraic

```
V A L U E S D E X P R E S S I O N
Q A C V B N M V A S D F U B H U I
W L R L K J G A C T U P B V M A L
E G G I L S P L A S H W S I K L V
R E C V A R T U M J U I T C V G N
T B N H T B O A S D F G I P O E U
Y R X O P N L T E R W T T Y T B E
U A V G T M K E U H B J U W Q R D
O I P O I U Y T F I J O T A X A G
P C O N S T A N T D F G E Z X C V
```

This Ron Howard movie was all wet.

______ ____ ______ _______ ______ ______

Holt Mathematics

<table><tr><td>LESSON
1-8</td><td>

Practice A
Translate Words Into Math
</td></tr></table>

Write as an algebraic expression.

1. the sum of *m* and 8

____

2. the product of 3 and *n*

____

3. 4 less than *x*

____

4. the quotient of a number and 12

____

5. 52 times a number

____

6. *w* less than 15

____

7. the sum of 13 and a number

____

8. the product of 5 and *p*, increased by 10

____

9. the sum of 15 divided by *b* and 6

____

10. 12 less than the amount *y* divided by 2

____

11. 26 increased by 12 times a number ______________________

12. the difference of 2 times a number and 6 ______________________

13. the product of *h* and 3, increased by 20 ______________________

14. 18 less than the product of a number and 4 ______________________

15. take away 32 from the product of 6 and a number ______________________

16. Used video games cost $25 each. Write an algebraic
expression to find the cost of *m* video games.   ______________________

17. Sal earned $740 for *n* weeks of work. Write an algebraic
expression for the amount he earned each week.   ______________________

18. At the end of the 2004–2005 NBA season, Reggie Miller
was the all-time leader in 3-point field goals made. He made
n more field goals than Dale Ellis. Dale Ellis made 1,719
3-pointers. Write an algebraic expression to find the number
of 3-pointers Reggie Miller made.   ______________________

19. The $2 bill has Thomas Jefferson on the front of it.
Write an algebraic expression to find out how much money
v bills with Thomas Jefferson on them would be worth. ______________________

Holt Mathematics

LESSON 1-8 Practice B
Translate Words Into Math

Write each phrase as an algebraic expression.

1. 125 decreased by a number

2. 359 more than z

3. the product of a number and 35

4. the quotient of 100 and w

5. twice a number, plus 27

6. 12 less than 15 times x

7. the product of e and 4, divided by 12

8. y less than 18 times 6

9. 48 more than the quotient of a number and 64 _________________

10. 500 less than the product of 4 and a number _________________

11. the quotient of p and 4, decreased by 320 _________________

12. 13 multiplied by the amount 60 minus w _________________

13. the quotient of 45 and the sum of c and 17 _________________

14. twice the sum of a number and 600 _________________

15. There are twice as many flute players as there are trumpet players in the band. If there are n flute players, write an algebraic expression to find out how many trumpet players there are. _________________

16. The Nile River is the longest river in the world at 4,160 miles. A group of explorers traveled along the entire Nile in x days. They traveled the same distance each day. Write an algebraic expression to find each day's distance. _________________

17. A slice of pizza has 290 calories, and a stalk of celery has 5 calories. Write an algebraic expression to find out how many calories there are in a slices of pizza and b stalks of celery. _________________

18. Grant pays 10¢ per minute plus $5 per month for telephone long distance. Write an algebraic expression for m minutes of long-distance calls in one month. _________________

Holt Mathematics

Practice C
Translate Words Into Math

Write each phrase as an algebraic expression.

1. the product of 6 and the square of a number _________________________

2. the square of the product of 6 and a number _________________________

3. 4 times the sum of a number and 6,008 _________________________

4. 200 less than half of a number _________________________

5. 3 times the difference of a number squared and 82 _________________________

6. 999 less than 45 increased by the product of a number

    and 85 _________________________

Write a verbal expression for each algebraic expression.

7. $2(4n)$ _________________________

8. $100 - \dfrac{54}{w}$ _________________________

9. $r^2 + 4r + 7$ _________________________

10. $\dfrac{45}{5s^2}$ _________________________

11. An albatross can sleep while flying 25 mi/h. An albatross
    flew 3 miles awake and another n hours asleep at
    25 mi/h. Write an algebraic expression to find the
    distance flown. _________________________

12. You have d dimes, q quarters, and n nickels.
    Write an algebraic expression to find the total
    amount of money. _________________________

13. Four out of every 10 homes in the United States
    have a dog. Write an algebraic expression to find
    out how many dogs there are in h homes. _________________________

14. A waitress who worked k hours earned $32 in
    tips. She gets an additional salary of $4.50 per
    hour. Write an algebraic expression to find the
    amount she earned. _________________________

Holt Mathematics

LESSON 1-8 Reteach
Translate Words Into Math

Use the operation clues in a word phrase to translate word phrases into algebraic expressions.

Addition	
add	plus
sum	more than
increased by	

Subtraction	
subtract	minus
difference	less than
decreased by	take away

Multiplication
times
multiplied by
product

Division
divided by
divided into
quotient

Write an algebraic expression for the difference of a number and 8.

1. What operation would you choose? _________________

2. Write an algebraic expression. _________________

Write an algebraic expression for 3 more than a number.

3. What operation would you choose? _________________

4. Write an algebraic expression. _________________

Write an algebraic expression for the quotient of a number and 15.

5. What operation would you choose? _________________

6. Write an algebraic expression. _________________

Write an algebraic expression.

7. the product of 12 and a number k _________________

8. a number d increased by 9 _________________

9. a number h divided by 4 _________________

Holt Mathematics

Challenge

What's My Equation?

Match each equation with the word problem it represents. Write the equation and corresponding letter. Then write the letter of the equation in the circle that has the problem number. Discover the message formed by the letters.

P $2k = 14$	**D** $\frac{p}{3} = 2.50$	**A** $\frac{m}{5} = 8$	**U** $3t = 21$
M $a - 18 = 33$	**S** $w + 7 = 25$	**T** $x + 7 = 22$	**H** $n - 12 = 37$

1. Tom has 7 more CDs than Rick does. If Tom has 22 CDs, how many does Rick have? _______________

2. Maria is 18 years younger than Kim. If Maria is 33, how old is Kim? _______________

3. Five friends went out for dinner. They shared the cost of the meal equally. If each person paid $8, what was the total cost of the meal? _______________

4. Max paid 3 times as much for a tape as his friend did. If Max paid $21, how much did his friend pay? _______________

5. Marisa and her two friends share a pizza. The cost of the pizza is shared equally among them. If each person pays $2.50, how much does the pizza cost? _______________

6. Lee has scored twice as many goals as Jiang has. If Lee's goal total is 14, how many goals has Jiang scored? _______________

7. Jamal sold 7 more magazine subscriptions than Wayne did. If Jamal sold 25 subscriptions, how many did Wayne sell? _______________

8. Shelly delivered 12 fewer newspapers this week than last week. If she delivered 37 papers this week, how many did she deliver last week? _______________

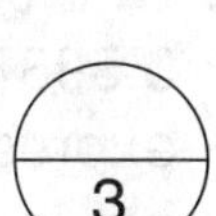

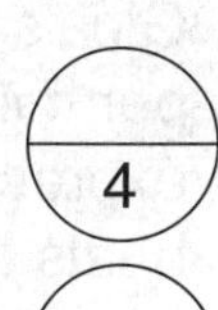

Holt Mathematics

<table>
<tr><td>LESSON
1-8</td><td></td></tr>
</table>

Problem Solving
Translate Words into Math

Write the correct answer.

1. Employers in the United States allocate *n* fewer vacation days than the 25 days given by the average Japanese employer. Write an algebraic expression to show the number of vacation days given U.S. workers.

2. There are 112 members in the Somerset Marching Band. They will march in *r* equal rows. Write an algebraic expression for the number of band members in each row.

3. A cup of cottage cheese has 26 grams of protein. Write an algebraic expression for the amount of protein in *s* cups of cottage cheese.

4. Every morning Sasha exercises for 20 minutes. She exercises *k* minutes every evening. Next week she will double her exercise time at night. Write an algebraic expression to show how long Sasha will exercise each day next week.

Choose the letter for the best answer.

5. One centimeter equals 0.3937 inches. Which algebraic expression shows how many inches are in *c* centimeters?

   A $0.3937 + c$

   B $0.3937 \div c$

   C $c \div 0.3937$

   D $0.3937c$

6. In 1957, the Soviet Union launched *Sputnik 1,* the first satellite to orbit Earth. It circled Earth's orbit every 1.6 hours for 92 days, then burned up. If the satellite traveled *m* miles per hour, which algebraic expression shows the length of the orbit?

   F $92m$

   G $1.6m$

   H $m \div 1.6$

   J $92 \div m$

7. Gina's heart rate is 70 beats per minute. Which algebraic expression shows the number of beats in *h* hours?

   A $70h$

   B $60h$

   C $4,200h$

   D $3,600h$

8. The Harris family went on vacation for *w* weeks and 3 days. Which algebraic expression shows the total number of days of their vacation?

   F $7w$

   G $3w$

   H $7w + 3$

   J $3w + 7$

Holt Mathematics

LESSON 1-8 Reading Strategies
Use a Graphic Organizer

Organizing word phrases for the four operations can help you write
algebraic expressions. Study the phrases and the key words
highlighted in this visual map.

$r + 12$	$5w$ or $5 \cdot w$
• **add** 12 to a number	• 5 **times** a number
• a number **plus** 12	• 5 **multiplied by** a number
• the **sum** of a number and 12	• the **product** of 5 and a number
• 12 **more than** a number	
• a number **increased by** 12	

Word Phrases for Algebraic Expressions

$n - 8$	$r \div 4$ or $\frac{r}{4}$
• **subtract** 8 from a number	• 4 **divided into** a number
• a number **minus** 8	• the **quotient** of a number with a divisor of 4
• 8 **less than** a number	• a number **divided by** 4
• a number **decreased by** 8	
• **take away** 8 from a number	

Write a word phrase for each algebraic expression.

1. $n - 15$ _______________________________

2. $(m + 12) - 3$ _______________________________

3. $2(y + 8)$ _______________________________

4. $5 + \dfrac{w}{3}$ _______________________________

Write an algebraic expression for each word phrase.

5. a number decreased by 9 _______________________________

6. the product of 15 and r _______________________________

7. the quotient of a number with a divisor of 5 _______________________________

8. three times the sum of n and 6 _______________________________

Holt Mathematics

LESSON 1-8 — Puzzles, Twisters & Teasers

Birds of a Feather!

Solve the crossword puzzle. Then use the letters in the shaded boxes to answer the riddle. You'll need to use some letters more than once.

Across

1. something that does not change

2. a number or symbol placed to the right of and above another number, symbol, or expression

6. putting together

8. the letter in a term

9. a number, a variable, or a product of numbers and variables

Down

1. the number in a term

3. the multiple expressed by an exponent

4. a symbol used for counting

5. group similar objects

7. similar

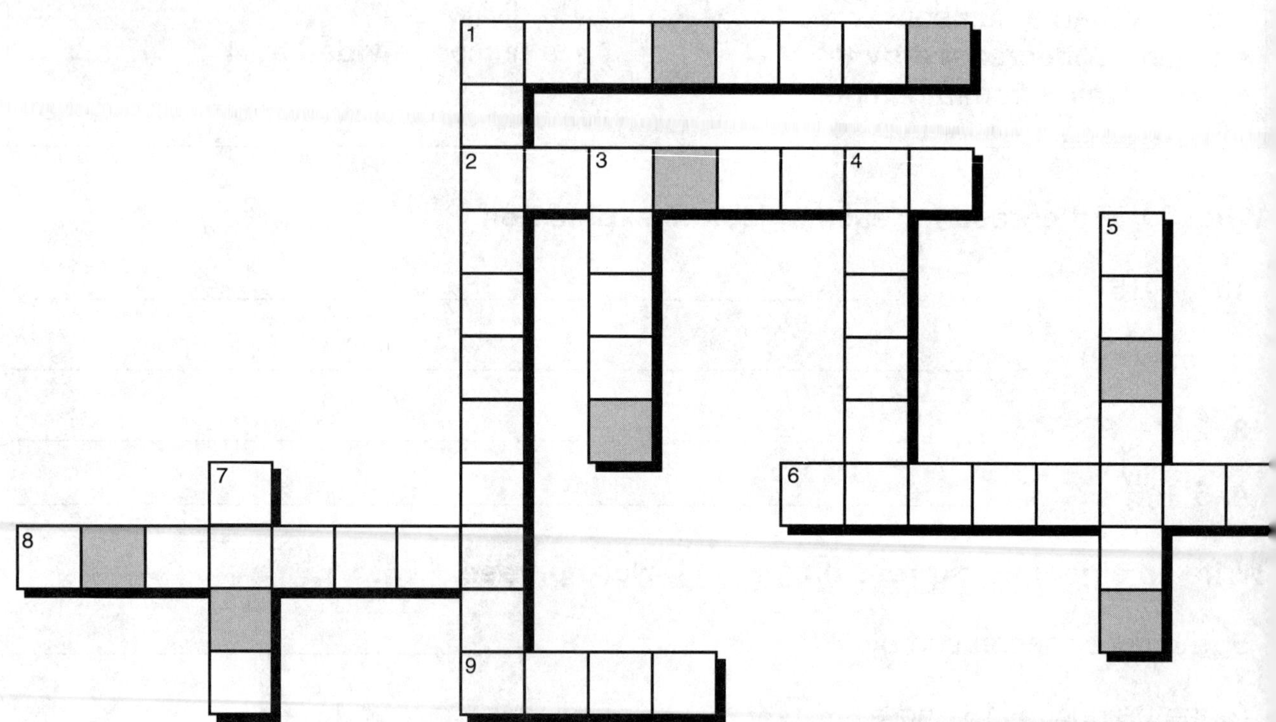

Where do birds invest their money?

In the ____ ____ ____ ____ ____   ____ ____ ____ ____ ____ ____.

Holt Mathematics

LESSON 1-9

Practice A
Simplifying Algebraic Expressions

Identify like terms in each list.

1. $6a$ b a 17 $4b$ 32 $17a$

____

2. x x^2 $3x$ 3 $3x^2$ 6

____

3. 2 $6z$ $6z^2$ z $17z$ z^2 3

____

4. m 8 $8m^2$ $8m$ m^2 $12m$ 18

____

5. $2p$ $22p$ $56q$ 12^2 q 34

____

6. d d^2 $15d^2$ $2d$ 4^2 $5d$ 44

____

Combine like terms.

7. $6p^2 + 3p^2$

8. $9x - 6x$

9. $a^2 + b^2 + 2a^2 + 5b^2$

10. $7h^2 + 3 - 2h^2 + 4$

11. $3x + 3y + x + y + z$

12. $5b + 5b + 6b^2 - 10 - 3b$

13. Find the perimeter of the rectangle. Combine like terms.

A $4x + 3y$

B $8x + 6y$

C $12xy$

D $4x^2 + 3y^2$

Holt Mathematics

LESSON 1-9
Practice B
Simplifying Algebraic Expressions

Identify like terms in each list.

1. $3a$ b^2 b^3 $4b^2$ 4 $5a$

____

2. x x^4 $4x$ $4x^2$ $4x^4$ $3x^2$

____

3. $6m$ $6m^2$ n^2 $2n$ 2 $4m$ $5n$

____

4. $12s$ $7s^4$ $9s$ s^2 5 $5s^4$ 2

____

Simplify. Justify your steps using the Commutative, Associative, and Distributive Properties when necessary.

5. $2p + 22q^2 - p$

6. $x^2 + 3x^2 - 4^2$

7. $n^4 + n^3 + 3n - n - n^3$

8. $4a + 4b + 2 - 2a + 5b - 1$

9. $32m^2 + 14n^2 - 12m^2 + 5n - 3$

10. $2h^2 + 3g - 2h^2 + 2^2 - 3 + 4g$

11. Write an expression for the perimeter of the figure at the right. Then simplify the expression.

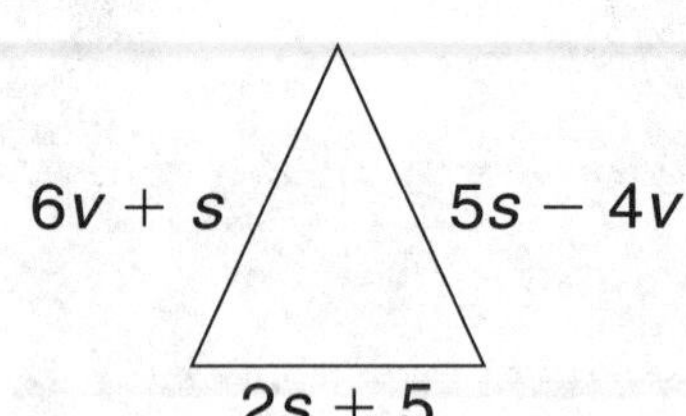

12. Write an expression for the combined perimeters of the figures at the right. Then simplify the expression.

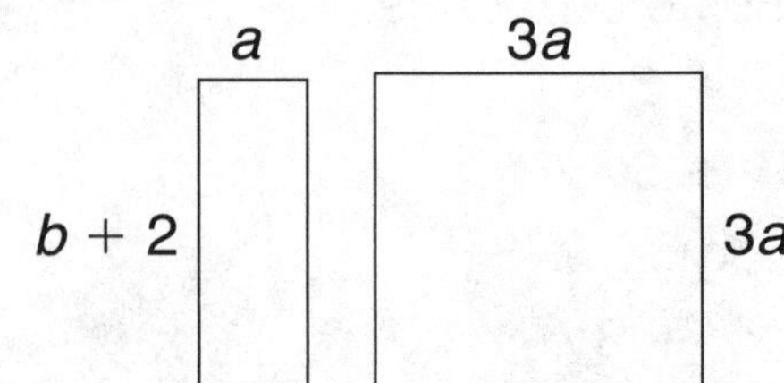

Holt Mathematics

Practice C
Simplifying Algebraic Expressions

Simplify each expression. Justify your steps using the Commutative, Associative, and Distributive Properties, when necessary.

1. $8k^2 + 4k - 3k^2 + 3^2 - k + 5$

2. $10x^3 + 5y^2 + 2xy - 4y^2 + 4xy - x^3$

____

____

3. $3a + 2b^2 + 6c + a - 2c + b^2 + c$

4. $12x^4 + 6x^2 + 5x^3 - x^2 + 2xy - 8x^4$

____

____

5. $9p^6 + q^2 + 6p + 5q^2 + 5p - 5q^2$

6. $h^2 + 4h + 4h^2 - h + 4 + h^2 + 7h$

____

____

7. Write an expression that has five terms and simplifies to $5m^3 + 4n$.

____

8. Write and simplify an expression for the perimeter of the figure to the right.

____

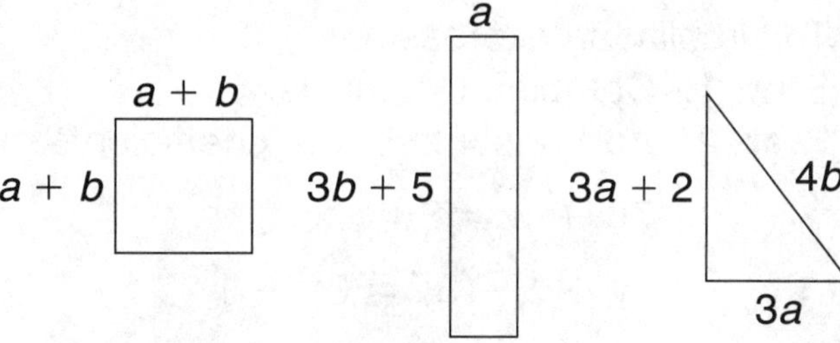

9. Write an expression to find the combined perimeters of the figures to the right. Then simplify the expression.

____

10. Jake scored *x* points in the first basketball game. He scored 2 fewer points in the next game. His teammate, Jack, scored *2y* points in the first game and 4 more than twice as many points in the next game. Write and simplify an expression for the total number of points scored by both players.

____

Holt Mathematics

Reteach

LESSON 1-9

Simplifying Algebraic Expressions

Look at the following expressions: $x = 1x$

$$x + x = 2x$$
$$x + x + x = 3x$$

The numbers 1, 2, and 3 are called **coefficients** of x.

Identify each coefficient.

1. $3n$ _______ **2.** $7y$ _______ **3.** m _______ **4.** 9 _______

An algebraic expression has terms that are separated by $+$ and $-$.
In the expression $2x + 5y$, the **terms** are $2x$ and $5y$.

Expression	Terms
$8x + 4y$	$8x$ and $4y$
$m - 3n$	m and $3n$
$4a^2 - 2b + a$	$4a^2$, $2b$, and a
$6d + 2p$	$6d$ and $2p$

Sometimes the terms of an expression can be combined.
Only **like terms** can be combined.

$7w + w$ like terms

$2x - 2y + 2$ unlike terms because x and y are different variables

$8e - 3e + 2e$ like terms

$5d + 25g$ unlike terms because d and g are different variables

To simplify an expression:
Step 1: Combine like terms.
Step 2: Add or subtract the coefficients of the variable.

$$7w + w = 8w$$

$$6y + 1 - 3y = 3y + 1$$

Simplify.

5. $y + 5y$ **6.** $9x - 4x$ **7.** $5s - 2s$ **8.** $3d + 7d$

____

9. $3b + b + 6$ **10.** $8a - a - 3$ **11.** $2p + 4p + r$ **12.** $9b - 8b + c$

____

Holt Mathematics

Challenge

LESSON 1-9

Matching Terms

Draw a line from each set of terms in Column A to its equivalent combination in Column B. Then circle each letter in Column B that does not have a matching term. Unscramble those letters to answer the riddle.

Column A	Column B
1. $2x + 7 + 5x - 4 - x$	A. $5y + 9x + 12$
2. $5 + 7x + 2x - 3 + 6$	B. $12y + 6x + 24$
3. $x + y + 4x - 3x + 2y + 3y$	C. 15
4. $3x^2 + 5x - 17 + 6x + 20$	D. $9x + 8$
5. $4x + x^2 + 12 - 4 + 2x$	E. 4
6. $12y + 12x + 12 - 6x + 12$	F. $6x + 3$
7. $12y + 4 + x - 7y + 8 + 8x$	G. $11x + y + 7$
8. $5x + x^2 + 2x + 5 - 4 - x^2$	H. $x^2 + 6x + 8$
9. $5x^2 + 8x + 7x^2 + 6x$	I. $4x$
10. $12x + 6 - 8x - 4x - 3 + 12$	J. $3x^2 + 11x + 3$
11. $5x + 4 - 3x + 5 + 2x - 9$	K. $3x + 2$
12. $4x + 2y + 8 - 3 - y - x$	L. $3x^2$
13. $4x + 5 + 7x + 2y + 2 - y$	M. $6x$
14. $2y + 2x + 8 - 6 + x - 2y$	N. $x^2 + 3x$
15. $4x + 6y + 6 + 7x + y$	O. $6x^2 + 6y + 1$
16. $3x^2 + 4x - 2x^2 - 3x + 2x$	P. $12x^2 + 14x$
17. $8x + 4 - 4 - 4x + x$	Q. $7x + 1$
18. $y + 5x + 6y + 9 - 6$	R. x^2
19. $x^2 + 3 + 2x^2 + 4 - 7$	S. $5x + 7y + 3$
20. $5y + 3 + 7x^2 - 2 - x^2 + y$	T. 0
	U. $2x + 6y$
	V. $3x + y + 5$
	W. $11x + 7y + 6$
	X. $5x$

Riddle: What can be a word, a number, a period of time, or a variable?

A _____ _____ _____ _____

Holt Mathematics

<table><tr><td>**LESSON**
1-9</td><td></td></tr></table>

Problem Solving
Simplifying Algebraic Expressions

Write the correct answer. Use the figures for Problems 1–3.

1. Figure 1 shows the length of each side of a garden. Write and simplify an expression for the perimeter of the garden.

   ____

2. Figure 2 is a square swimming pool. Write and simplify an expression for the perimeter of the pool.

   ____

3. Write and simplify an expression for the combined perimeter of the garden and the pool.

   ____

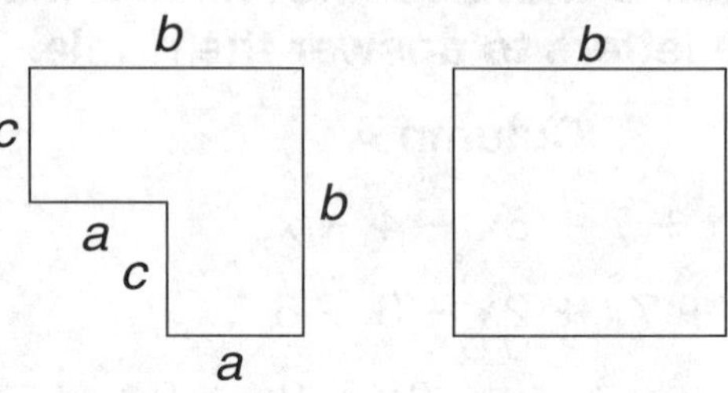

4. The Pantheon in Rome has n granite columns in each of 3 rows. Write and simplify an addition expression to show the number of columns. Then evaluate the expression for $n = 8$.

   ____

Choose the letter for the best answer.

5. Which is an expression that shows the earnings of a telemarketer who worked for 23 hours at a salary of d dollars per hour?

   A $d + 23$ **C** $d \div 23$

   B $23d$ **D** $23 \div d$

6. The minimum wage set in 1997 was $5.15 per hour. Evaluate the expression $40h$ where $h = \$5.15$ to find a worker's weekly salary.

   F $20.60   **H** $515.00

   G $200   **J** $206.00

7. What is the perimeter of a triangle with sides the following lengths: $2a + 4c$, $3c + 7$, and $6a - 4$. Simplify the expression.

   A $8a + 11c$

   B $6a + 7c + 3$

   C $8a + 7c + 3$

   D $8a + 7c + 11$

8. A hexagon is a 6-sided figure. Find the perimeter of a hexagon where all of the sides are the same length and the expression $x + y$ represents the length of a side. Simplify the expression.

   F $6x + 6y$

   G $6 + x + y$

   H $6x + y$

   J $6xy$

Holt Mathematics

Reading Strategies
Organization Patterns

An algebraic expression is made up of parts called **terms.**

constants	variables	constants and variables
3.2 $\frac{1}{2}$ 12	m s x	$\frac{n}{2}$ $\frac{2}{3}y$ $4x\ 3m^2$

A **coefficient** is a value multiplied by a variable.

Term	Value of Coefficient	Meaning
$7x$	7	$7 \bullet x$
y	1	$1 \bullet y$
$\frac{x}{3}$	$\frac{1}{3}$	$\frac{1}{3} \bullet x$

This expression has 6 terms:

Term   Term   Term   Term   Term   Term

$2x \quad + \quad 5b \quad + \quad 7 \quad - \quad b \quad + \quad 3x \quad + \quad 2x^2$

- Reorganize the terms: $2x + 3x + 5b - b + 7 + 2x^2$.
- Combine like terms: $5x + 4b + 7 + 2x^2$.

Answer each question.

1. How many terms are there in this expression:
$6b + b^2 + 5 + 2b - 3f$?

____

2. $6b$ and b^2 are unlike terms. Explain why.

____

3. How many terms are there in this expression:
$5a^2 + 6b + a^3 + 2a^2 - 3b - 2$?

____

4. Reorganize these terms so like terms are next to each other.

____

Holt Mathematics

Puzzles, Twisters & Teasers

LESSON 1-9

In Other Words...

Write each verbal expression as an algebraic expression. Then use the answer key to solve the riddle.

1. the product of 20 and *t* _______________

2. the sum of 4 times a number and 2 _______________

3. the product of 7 and *p* _______________

4. the sum of six times a number and 1 _______________

5. the sum of 5 and a number _______________

6. the quotient of a number and 8 _______________

7. *m* plus 6 _______________

8. *t* less than 23 _______________

9. the quotient of 100 and the amount 6 plus *w* _______________

Answer Key

+	−	×	÷
S H L B	N	G A	C I

What's worse than raining cats and dogs?

___ ___ ___ ___ ___ ___ ___
+   ×   ÷   +   ÷   −   ×

___ ___ ___ ___
÷   ×   +   +

74

Holt Mathematics

Practice A
Equations and Their Solutions

Tell if the value of the variable is a solution for the equation.

1. $n = 22$ for $n + 5 = 27$ **2.** $a = 43$ for $72 - a = 30$ **3.** $g = 31$ for $19 + g = 40$

_______________   _______________   _______________

4. $z = 87$ for $z - 22 = 65$ **5.** $f = 15$ for $46 + f = 61$ **6.** $h = 53$ for $h - 24 = 77$

_______________   _______________   _______________

7. $p = 4$ for $5p = 20$ **8.** $k = 48$ for $k \div 8 = 8$ **9.** $m = 12$ for $48 \div m = 4$

_______________   _______________   _______________

10. $x = 8$ for $32x = 264$ **11.** $v = 5$ for $\dfrac{90}{v} = 15$ **12.** $s = 7$ for $16s = 112$

_______________   _______________   _______________

13. $m = 2$ for $3m + 4 = 10$ **14.** $y = 6$ for $3y - 4 = 18$ **15.** $r = 10$ for $10r - 10 = 90$

_______________   _______________   _______________

16. In the United States in 2005, there were 68 endangered species that were mammals. There were 9 more endangered species that were birds. Were there 59 or 77 endangered species of birds in the United States?

17. At the Bike and Blade Shop, mountain bikes are on sale for $349. This is $30 more than a racing bike costs. Does the racing bike cost $319 or $379?

_______________   _______________

18. Which problem situation best matches the equation $2c + 10 = 350$?

Situation A: Austin paid $10 for a new video game console. This is $350 more than 2 times the cost of a used console. How much does a used console cost?

Situation B: Austin paid $350 for a new video game console. This is $10 more than 2 times the cost of a used console. How much does a used console cost?

____

Holt Mathematics

Practice B
Equations and Their Solutions

Determine whether the given value of the variable is a solution.

1. $a = 4$ for $12 - a = 6$ **2.** $m = 37$ for $23 + m = 60$ **3.** $x = 6$ for $54 = 9x$

_______________   _______________   _______________

4. $g = 96$ for $\frac{g}{4} = 32$ **5.** $n = 28$ for $n + 44 = 72$ **6.** $j = 6$ for $84 \div j = 12$

_______________   _______________   _______________

7. $k = 24$ for $3k = 6$ **8.** $m = 3$ for $42 = m + 39$ **9.** $y = 8$ for $8y + 6 = 70$

_______________   _______________   _______________

10. $s = 5$ for $18 = 3s - 3$ **11.** $k = 7$ for $23 - k = 30$ **12.** $v = 12$ for $84 = 7v$

_______________   _______________   _______________

13 $c = 15$ for $45 - 2c = 15$ **14.** $x = 10$ for $x + 25 - 2x + 4 = 19$

_______________   _______________

15. $e = 6$ for $42 = 51 - e$ **16.** $p = 15$ for $19 = p - 4$

_______________   _______________

17. Jason and Maya have their own web sites on the Internet. As of last week, Jason's web site had 2,426 visitors. This is twice as many visitors as Maya had. Did Maya have 1,213 visitors or 4,852 visitors to her web site?

____

18. Which problem situation best matches the equation $3c - 5 = 31$?

Situation A: Rachel had a coupon for $5 off the cost of her order. She ordered 3 large pizzas that each cost the same amount and paid a total of $31. What was the cost of each pizza?

Situation B: Rachel had a coupon for $31 off the cost of her order. She ordered 5 large pizzas that each cost the same amount and paid a total of $3. What was the cost of each pizza?

____

Holt Mathematics

Practice C
Equations and Their Solutions

Determine whether the given value of the variable is a solution.

1. $a = 15$ for $75 \div a = 5$

2. $x = 90$ for $x \div 9 = 100 - x$

3. $d = 8$ for $875 = 909 - 4d$

4. $x = 32$ for $2x - 25 + x - 70 = 1$

5. $e = 2$ for $e^3 - e^2 = 6e - 8$

6. $b = 4$ for $b^2 + 2b - 3 = 27$

7. $d = 12$ for $4d - 24 - 12 = 0$

8. $r = 9$ for $4r^2 - 19 - 5r = 340$

9. $p = 25$ for $\dfrac{4p}{5} + 2p + 10 = 80$

10. $t = 18$ for $\dfrac{t}{6} + \dfrac{54}{t} = 3$

11. In 1993, there were about 34,000,000 cell phone subscribers in the United States. This is about 125,000,000 fewer subscribers than there were by the year 2003. The equation $34{,}000{,}000 = s - 125{,}000{,}000$ can be used to represent the number of cell phone subscribers in 2003. Were there 91,000,000 or 159,000,000 cell phone subscribers in the United States in 2003?

____

12. Which problem situation best matches the equation $50n + 15 = 150$?

Situation A: Hector paid a commission of $150 to buy 15 shares of an Internet stock. Altogether he paid $50. How much did each share of stock cost?

Situation B: Hector paid a commission of $15 to buy 50 shares of an Internet stock. Altogether he paid $150. How much did each share of stock cost?

____

Holt Mathematics

Reteach
Equations and Their Solutions

Number sentences that contain an equal sign (=) are called **equations.**

Equations may be true, or they may be false.

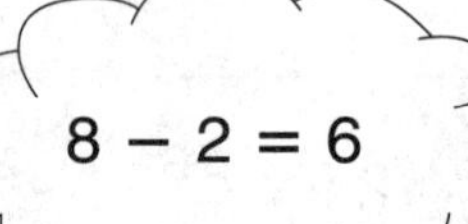

True	False
$3 + 4 = 7$	$3 + 1 = 7$
$8 - 6 = 2$	$8 - 2 = 5$

An equation may contain a variable.

variable → $\quad x + 4 = 6$

Whether this equation is true or false depends on the value of x.

You can decide if a number is a *solution* of an equation. Substitute the number for the variable in the equation. If the equation is a true equation, then the number is the **solution.**

Equation: $x + 4 = 6$

Is 2 a solution?

$$x + 4 = 6$$

Substitute 2 for x

$$2 + 4 \overset{?}{=} 6$$
$$6 \overset{?}{=} 6 \quad \text{True}$$

2 is a solution of $x + 4 = 6$.

Is 3 a solution?

$$x + 4 = 6$$

Substitute 3 for x

$$3 + 4 \overset{?}{=} 6$$
$$7 \overset{?}{=} 6 \quad \text{False}$$

3 is not a solution of $x + 4 = 6$.

Tell if the number is a solution.

1. Is 3 a solution of $y + 3 = 9$?

____

2. Is 4 a solution of $n + 6 = 10$?

____

3. Is 2 a solution of $w - 1 = 1$?

____

4. Is 1 a solution of $x + 50 = 49$?

____

5. Is 6 a solution of $c + 23 = 30$?

____

6. Is 9 a solution of $v - 9 = 0$?

____

7. Is 20 a solution of $t - 17 = 3$?

____

8. Is 16 a solution of $12 + a = 24$?

____

9. Is 25 a solution of $38 - m = 13$?

____

10. Is 8 a solution of $15 = e + 5$?

____

Holt Mathematics

Name ____ Date ____________ Class ____________

Challenge
The Solution Is BINGO!

Find the solution to each problem or equation. Cross it out on the board below to get BINGO!

1. Is 37, 47, or 67 a solution for
   $52 = n + 15$?

2. Is 14, 17, or 21 a solution for
   $8y - 7 = 129$?

3. Is 14, 22, or 24 a solution for
   $132 - (4x - 5) = 81$?

4. Is 12, 15, or 18 a solution for
   $3(60 - s) - 2s = 105$?

5. Garret scored 18 points in his last basketball game, which is 6 fewer points than Vince scored. The equation $18 = p - 6$ can be used to represent Vince's points. Did Vince score 12, 24, or 28 points?

6. In 3 years, Sarah's sister will be twice as old as Sarah. If Sarah is now 3 years old, will her sister be 6, 9, or 12 years old in 3 years?

7. The highest recorded temperature in Alaska was 100°F in 1915. This was 56 years before the lowest recorded temperature of −80°F. Was the lowest recorded temperature in 1856, 1956, or 1971?

8. In 2002, Florida had 3,314 public schools. This was 155 more public schools than 3 times the number of public schools in South Carolina during the same year. Did South Carolina have 1,053, 2,849, or 3,159 public schools in 2002?

B	I	N	G	O
17	67	24	37	1971
6	24	47	3,159	22
2,849	1956	FREE	18	14
12	9	16	1,053	1856
28	23	21	19	15

Holt Mathematics

LESSON 1-10 Problem Solving
Equations and Their Solutions

Write the correct answer.

1. The jet airplane was invented in 1939. This is 12 years after the first television was invented. Was television invented in 1927 or 1951?

2. There are three times as many students in the high school as in the junior high school, which has 330 students. Does the high school have 990 students or 110 students?

3. The frigate bird has been recorded at speeds up to 95 mi/h. The only faster bird ever recorded was the spine-tailed swift at 11 mi/h faster. Was the speed of the spine-tailed swift 84 mi/h or 106 mi/h?

4. As of 2004, there were 20.5 million Internet users in Canada. This is 6.6 million more Internet users than there were in Mexico. Were there 27.1 million or 13.9 million Internet users in Mexico?

Choose the letter for the best answer.

5. In the United States, the average school year is 180 days. This is 71 days less than the average school year in China. What is the average school year in China?

   A 251 days

   B 109 days

   C 151 days

   D 271 days

6. The longest suspension bridge in the world is the Akashi Kaikyo Bridge in Japan. Its main span is 1,290 feet longer than a mile. A mile is 5,280 feet. How long is the Akashi Kaikyo bridge?

   F 3,990 feet

   G 6,400 feet

   H 4,049 feet

   J 6,570 feet

7. *Ornithomimus* stood about 6 feet tall and was the fastest dinosaur at a speed of about 50 mi/h. The largest dinosaur, *Seismosaurus,* was 20 times as tall. How tall was *Seismosaurus?*

   A 12 feet

   B 70 feet

   C 120 feet

   D 26 feet

8. Milton collects sports trading cards. He has 80 baseball cards. He has half as many basketball cards as football cards. He has 20 more hockey cards than basketball cards and half as many football cards as baseball cards. How many hockey cards does he have?

   F 20 hockey cards

   G 40 hockey cards

   H 60 hockey cards

   J 80 hockey cards

Holt Mathematics

Reading Strategies
Use a Visual Model

The expressions on both sides of the equal sign are equal in
an **equation.**
You can read an equation in math much like you read a sentence.
When you see an equal sign, you read, **"is the same as."**
Balanced scales can help you picture an equation.

Mark has 45 baseball cards.　→　Mark's cards are equal to
This is 8 more than his sister Kathy.　　Kathy's cards + 8.

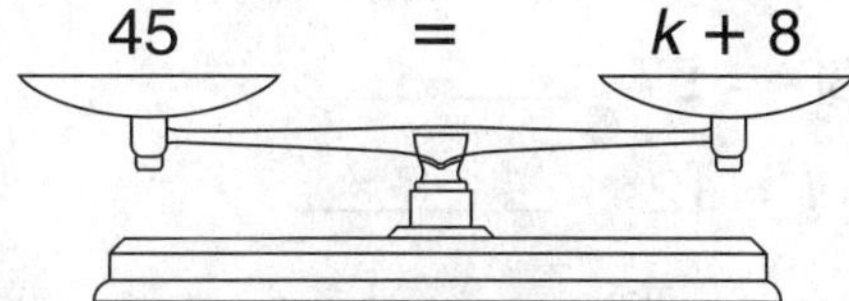

The balanced scales illustrate this equation. →

The value of the variable that makes the equation a true sentence is
called the **solution.**

Substitute 40 for k: $45 = k + 8$　　　　Substitute 37 for k: $45 = k + 8$
$45 \neq 40 + 8$　　　　　　　　　　　　　　$45 = 37 + 8$
40 is not a solution.　　　　　　　　　　**37 is the solution.**

Answer each question.

1. What is an equation?

____

____

2. What is the solution of an equation?

____

____

3. Is $3 + 28 \neq 32$ an equation? Why or why not.

____

4. Is $n = 12$ a solution for the equation $43 = n + 31$?
   Why or why not?

____

Determine whether the given value of the variable is a solution.

5. $d + 37 = 54$ for $d = 23$ _______________________________________

6. $t - 23 = 42$ for $t = 56$ _______________________________________

Holt Mathematics

Puzzles, Twisters & Teasers

1-10 *The Answer Popped Right into My Head!*

Find the solution for each equation below. Write the letter on the line above the correct answer at the bottom of the page to solve the riddle.

P $18 = s - 7$ ________

M $x + 3 = 10$ ________

O $12 = t + 9$ ________

Y $s - 38 = 57$ ________

R $16 - j = 12$ ________

I $16 = 34 - m$ ________

U $48 = x + 12$ ________

S $17 + k = 40$ ________

N $24 = 34 - n$ ________

B $p + 18 = 29$ ________

A $47 = v - 6$ ________

G $94 = c + 6$ ________

H $82 = j + 9$ ________

E $a - 15 = 17$ ________

T $15 = k - 9$ ________

What did one firecracker say to the other?

____ ____   ____ ___ ____   ____ ____
 7   95   25   3   25   18   23

____ ____ ____ ____ ____ ___   ____ ____ ____ ____
11   18   88   88   32   4   24   73   53   10

____ ___ ____ ___   ____ ___ ____ .
95   3   36   4   25   3   25

Holt Mathematics

Practice A
Solving Equations by Adding or Subtracting

Match each equation in Column A with its correct solution in Column B.

Column A	Column B	Column A	Column B
1. $n - 16 = 8$	A. $n = 12$	10. $x - 12 = 13$	L. $x = 14$
2. $5 = n - 7$	B. $n = 13$	11. $x + 8 = 40$	M. $x = 17$
3. $12 + n = 25$	C. $n = 17$	12. $34 = 16 + x$	N. $x = 18$
4. $n - 17 = 11$	D. $n = 24$	13. $x + 5 = 19$	P. $x = 25$
5. $n + 18 = 35$	E. $n = 27$	14. $4 + x = 52$	Q. $x = 32$
6. $7 = n - 28$	F. $n = 28$	15. $12 + x = 50$	R. $x = 33$
7. $n - 12 = 40$	G. $n = 35$	16. $15 = x - 2$	S. $x = 38$
8. $24 = n - 25$	H. $n = 49$	17. $52 = x + 9$	T. $x = 43$
9. $46 = n + 19$	J. $n = 52$	18. $x - 11 = 22$	U. $x = 48$

19. Chris has 55 baseball trading cards. He has 17 more cards than his sister Sara has. Write and solve an equation to find how many trading cards Sara has.

____

20. In 2000, Sammy Sosa hit 50 home runs. His home run total was 23 runs fewer than the number of home runs that Barry Bonds hit the next year. Write and solve an equation to find how many home runs Barry Bonds hit in 2001.

____

Holt Mathematics

Practice B
Solving Equations by Adding or Subtracting

Solve each equation. Check your answer.

1. $33 = y - 44$ **2.** $r - 32 = 77$ **3.** $125 = x - 29$

_______________   _______________   _______________

4. $k + 18 = 25$ **5.** $589 + x = 700$ **6.** $96 = 56 + t$

_______________   _______________   _______________

7. $a - 9 = 57$ **8.** $b - 49 = 254$ **9.** $987 = f - 11$

_______________   _______________   _______________

10. $32 + d = 1{,}400$ **11.** $w - 24 = 90$ **12.** $95 = g - 340$

_______________   _______________   _______________

13. $e - 35 = 59$ **14.** $84 = v + 30$ **15.** $h + 15 = 81$

_______________   _______________   _______________

16. $110 = a + 25$ **17.** $45 + c = 91$ **18.** $p - 29 = 78$

_______________   _______________   _______________

19. $56 - r = 8$ **20.** $39 = z + 8$ **21.** $93 + g = 117$

_______________   _______________   _______________

22. The Morales family is driving from Philadelphia to Boston. So far, they have driven 167 miles. This is 129 miles less than the total distance they must travel. How many miles is Philadelphia from Boston?

____

23. Ron has $1,230 in his savings account. This is $400 more than he needs to buy a new big screen TV. Write and solve an equation to find out how much the TV costs.

____

Holt Mathematics

<table>
<tr><td>LESSON</td></tr>
<tr><td>1-11</td></tr>
</table>

Practice C
Solving Equations by Adding or Subtracting

Solve each equation. Check your answer.

1. $b - 32 = 15$ **2.** $e - 43 = 121$ **3.** $601 = x - 24$

_______________   _______________   _______________

4. $m + 45 = 123$ **5.** $314 + z = 350$ **6.** $840 = 45 + f$

_______________   _______________   _______________

7. $d - 67 = 23$ **8.** $w + 233 = 319$ **9.** $91 = x + 52$

_______________   _______________   _______________

10. $150 + y = 879$ **11.** $k - 32 = 217$ **12.** $408 = s - 129$

_______________   _______________   _______________

13. $108 = j - 24$ **14.** $1{,}204 = w + 389$ **15.** $p - 167 = 321$

_______________   _______________   _______________

16. In 2003, *USA Today* was the leading U.S. daily newspaper, with a circulation of 2,154,539. This was 63,477 more than the second leading daily newspaper, the *Wall Street Journal*. Write and solve an equation to find the circulation of the *Wall Street Journal* in 2003.

____

17. Mrs. Baker's class has been raising money for a local charity. At the end of last week, they had collected $238. By the end of this week, they had a total of $419. Write and solve an equation to find the amount collected this week.

____

18. Andy Green broke the sound barrier on land on October 15, 1997, in Black Rock Desert, Nevada. The speed of sound was recorded at about 751 mi/h. This was about 12 mi/h less than the land speed record that he set that day. Write and solve an equation to find Andy Green's land speed record.

____

Holt Mathematics

Name _________________________________ Date __________ Class __________

 Solving Equations by Adding or Subtracting

Solving an equation is like balancing a scale. If you add the same weight to both sides of a balanced scale, the scale will remain balanced. You can use this same idea to solve an equation.

Think of the equation $x - 7 = 12$ as a balanced scale. The equal sign keeps the balance.

$$x - 7 = 12$$

$\boxed{-7 + 7 = 0}$ $x - 7 + \mathbf{7} = 12 + \mathbf{7}$ Add 7 to both sides.

$$x + 0 = 19 \quad \text{Combine like terms.}$$
$$x = 19$$

When you solve an equation, the idea is to get the variable by itself. What you do to one side of the equation, you must do to the other side.

• To solve a subtraction equation, use addition.

• To solve an addition equation, use subtraction.

Solve and check: $y + 8 = 14$.

$$y + 8 = 14$$

$\boxed{+8 - 8 = 0}$ $y + 8 - \mathbf{8} = 14 - \mathbf{8}$ Subtract 8 from both sides.

$$y + 0 = 6 \quad \text{Combine like terms.}$$
$$y = 6$$

Check: $y + 8 = 14$ To check, substitute 6 for *y*.

$$6 + 8 \overset{?}{=} 14$$
$$14 \overset{?}{=} 14 \quad ✔$$

A true sentence, $14 = 14$, means the solution is correct.

Solve and check.

1. $x - 2 = 8$ **2.** $b + 5 = 11$

    $x - 2 + \underline{\;\;\;} = 8 + \underline{\;\;\;}$ $b + 5 - \underline{\;\;\;} = 11 - \underline{\;\;\;}$

       $x - 0 = \underline{\;\;\;}$ $b + 0 = \underline{\;\;\;}$

3. $n + 8 = 11$ **4.** $y - 6 = 2$ **5.** $a - 9 = 4$ **6.** $m + 2 = 18$

_______________   _______________   _______________   _______________

Holt Mathematics

Challenge
Equation Maker

Use each term once to make up one addition and one subtraction equation, then solve the equations.

1. *m, n*, 12, 6, 54, 9

____

____

2. *x, y*, 7, 15, 32, 45

____

____

3. *p, q*, 19, 44, 72, 8

____

____

4. *a, b*, 67, 102, 6, 8

____

____

5. *c, d*, 11, 12, 18, 35

____

____

6. *s, t*, 115, 123, 32, 0

____

____

7. *w, y*, 1, 2, 3, 4

____

____

8. *n, p*, 6, 22, 99, 400

____

____

9. *e, f*, 52, 4, 75, 18

____

____

10. *g, h*, 61, 88, 94, 117

____

____

11. *k, l*, 302, 54, 115, 79

____

____

12. *r, s*, 90, 14, 71, 15

____

____

13. *u, v*, 8, 12, 37, 44

____

____

14. *x, y*, 198, 0, 231, 4

____

____

Holt Mathematics

Problem Solving

LESSON 1-11 *Solving Equations by Adding or Subtracting*

Write the correct answer.

1. In an online poll, 1,927 people voted for Coach as the best job at the Super Bowl. The job of Announcer received 8,055 more votes. Write and solve an equation to find how many votes the job of Announcer received.

2. In 2005, the largest bank in the world was UBS, Switzerland, with $1,533 billion in assets. This was $49 billion more than the largest bank in the United States, Citigroup. Write and solve an equation to find Citigroup's assets.

3. The two smallest countries in the world are Vatican City and Monaco. Vatican City is 1.37 square kilometers smaller than Monaco, which is 1.81 square kilometers in area. What is the area of Vatican City?

4. The Library of Congress is the largest library in the world. It has 29 million books, which is 10 million more than the National Library of Canada has. How many books does the National Library of Canada have?

Choose the letter for the best answer.

5. The first track on Sean's new CD has been playing for 55 seconds. This is 42 seconds less than the time of the entire first track. How long is the first track on this CD?

   A 37 seconds   **C** 97 seconds

   B 63 seconds   **D** 93 seconds

6. There are 45 students on the school football team. This is 13 more than the number of students on the basketball team. How many students are on the basketball team?

   F 58 students   **H** 32 students

   G 48 students   **J** 42 students

7. A used mountain bike costs $79.95. This is $120 less than the cost of a new one. If c is the cost of the new bike, which equation can you use to find the cost of a new bike?

   A $79.95 = c + 120$

   B $120 = 79.95 - c$

   C $79.95 = c - 120$

   D $120 = 79.95 + c$

8. The goal of the School Bake Sale is to raise $125 more than last year's sale. Last year the Bake Sale raised $320. If it reaches its goal, how much will the Bake Sale raise this year?

   F $445

   G $195

   H $525

   J $425

Holt Mathematics

Reading Strategies
LESSON 1-11 *Follow a Procedure*

In order to solve an equation, you must find the **solution.** The solution is the value that makes the equation true. To solve an equation, you need to get the variable by itself on one side of the equal sign.

- If you have an addition equation, you must subtract to get the variable by itself.

- If you have a subtraction equation, you must add to get the variable by itself.

Example:

$z + 12 = 32$ ◄— To get z by itself, subtract 12.

$z + 12 - \mathbf{12} = 32 - \mathbf{12}$ ◄— Rewrite the equation to show that 12 is subtracted from both sides.

$z = 20$ ◄— This is the solution after subtracting 12 from both sides.

Check by using 12 in place of z.

$20 + 12 \stackrel{?}{=} 32$

$32 = 32$, so $z = 20$ is the correct solution.

Example:

$27 = x - 8$ ◄— To get x by itself, add 8.

$27 + \mathbf{8} = x - 8 + \mathbf{8}$ ◄— Rewrite the equation to show that 8 is added to both sides.

$35 = x$ ◄— This is the solution after adding 8 to both sides.

Check by using 35 in place of x.

$27 \stackrel{?}{=} 35 - 8$

$27 = 27$, so $x = 35$ is the correct solution.

Use $m + 17 = 43$ for Exercises 1–4.

1. What operation is shown in this equation? _____________________________

2. What operation will you use to get m by itself? _____________________________

3. Rewrite the equation showing subtracting from both sides of the equation. _____________________________

4. What is the value of m? _____________________________

Holt Mathematics

Puzzles, Twisters & Teasers

LESSON 1-11 *Clean Solutions!*

Find the solution for each equation below. Write the letter of each variable on the line above the correct answer at the bottom of the page to solve the riddle.

1. $y - 35 = 17$ ____________

2. $h - 40 = 26$ ____________

3. $a + 16 = 43$ ____________

4. $110 = e + 66$ ____________

5. $97 = w - 44$ ____________

6. $n - 8 = 3$ ____________

7. $356 = g - 218$ ____________

8. $652 + t = 800$ ____________

9. $16 = o - 124$ ____________

10. $63 + m = 903$ ____________

11. $k + 18 = 98$ ____________

12. $d - 27 = 54$ ____________

13. $c - 50 = 23$ ____________

14. $l - 62 = 937$ ____________

Why did the robber take a shower?

____ ____   ____ ____ ____ ____ ____ ____   ____ ____
66   44   141 27 11   148 44   81   148 140

____ ____ ____ ____   ____ ____   ____ ____ ____ ____ ____
840 27   80   44   27   73   999 44   27   11

____ ____ ____   ____ ____ ____ ____ .
574 44   148   27   141 27   52

Holt Mathematics

Practice A
Solving Equations by Multiplying or Dividing

Solve.

1. $16 = n \div 2$

2. $\dfrac{e}{10} = 8$

3. $25 = \dfrac{x}{6}$

____

____

____

4. $18 = \dfrac{d}{3}$

5. $a \div 12 = 7$

6. $30 = b \div 4$

____

____

____

Solve and check.

7. $7w = 49$

8. $75 = 3x$

9. $60 = 12p$

____

____

____

10. $77 = 11m$

11. $4h = 48$

12. $9y = 54$

____

____

____

13. $2x = 30$

14. $45 = 5s$

15. $6z = 42$

____

____

____

16. The Fruit Stand charges $0.50 each for navel oranges. Kareem paid $4.00 for a large bag of navel oranges. How many did he buy?

____

17. Jenny can type at a speed of 80 words per minute. It took her 20 minutes to type a report. How many words was the report?

____

18. At the local gas station, regular unleaded gasoline is priced at $1.10 per gallon. If it cost $16.50 to fill a car's gas tank, how many gallons of gasoline did the tank hold?

____

Holt Mathematics

Practice B

LESSON 1-12 *Solving Equations by Multiplying or Dividing*

Solve each equation. Check your answer.

1. $68 = \dfrac{r}{4}$

2. $k \div 24 = 85$

3. $255 = \dfrac{x}{4}$

____

4. $42 = w \div 18$

5. $\dfrac{a}{15} = 22$

6. $82 = b \div 5$

____

7. $\dfrac{c}{7} = 9$

8. $28 = z \div 3$

9. $\dfrac{y}{12} = 10$

____

Solve each equation. Check your answer.

10. $52w = 364$

11. $41x = 492$

12. $410 = 82p$

____

13. $35d = 735$

14. $195 = 65h$

15. $4k = 140$

____

16. $110 = 5e$

17. $27a = 216$

18. $96 = 12n$

____

19. Ashley earns $5.50 per hour babysitting. She wants to buy a CD player that costs $71.50, including tax. How many hours will she need to work to earn the money for the CD player?

____

20. A cat can jump the height of up to 5 times the length of its tail. How high can a cat jump if its tail is 13 inches long?

____

Holt Mathematics

Practice C
Solving Equations by Multiplying or Dividing

Solve each equation. Check your answer.

1. $765 = \dfrac{n}{12}$

2. $\dfrac{m}{9} = 26$

3. $\dfrac{a}{12} = 14$

4. $3g = 165$

5. $308 = 44b$

6. $27e = 405$

Translate each sentence into an equation. Then solve the equation.

7. The product of a number w and 145 is 725. _______________________

8. The quotient of a number f and 21 is 14. _______________________

9. A number b times 23 equals 253. _______________________

10. A number k divided by 15 equals 47. _______________________

11. A number n multiplied by 12 is 84. _______________________

12. Ten divided into a number m equals 54. _______________________

13. About three tons of ore must be mined and processed to produce a single ounce of gold. How many tons of ore are required to produce a pound of gold?

14. It costs about \$2,931 per hour to operate a Boeing 757 airplane. Find the cost to operate a Boeing 757 during a 5-hour flight.

15. Each person in the United States eats an average of 23 quarts of ice cream per year. At this rate, the seventh-grade class will eat about 1,794 quarts of ice cream this year. How many students are in the seventh-grade class?

16. Jumbo shrimp sell for \$14.99 per pound in a local supermarket. One customer spent \$74.95 on shrimp for a dinner party. How much shrimp did this customer purchase for the party?

Holt Mathematics

<table><tr><td>LESSON
1-12</td><td># Reteach
Solving Equations by Multiplying or Dividing</td></tr></table>

When you solve an equation, you must get the variable by itself.
Remember, what you do to one side of an equation, you must do to
the other side.

- To solve a division equation, multiply both sides of the equation by
  the same number.

Solve and check: $\frac{a}{3} = 4$.

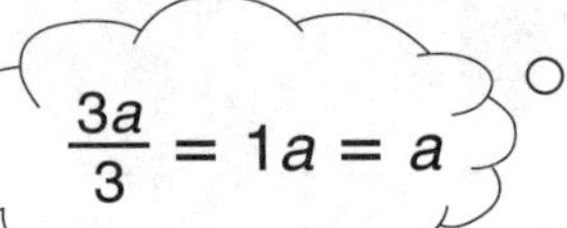

$$\frac{a}{3} = 4$$

$$(3)\,\frac{a}{3} = 4(3)$$

$$a = 12$$

Multiply to solve a division equation.

Check: $\frac{a}{3} = 4$

Replace the variable with the solution.

$$\frac{12}{3} \stackrel{?}{=} 4$$

$$4 \stackrel{?}{=} 4 \quad ✔$$

A true sentence means the solution is correct.

Solve and check.

1. $\frac{x}{6} = 3$ **2.** $\frac{c}{8} = 8$ **3.** $\frac{c}{10} = 7$ **4.** $\frac{n}{3} = 12$

_______________ _______________ _______________ _______________

- To solve a multiplication equation, divide both sides of the equation by
  the same number.

Solve and check: $5k = 30$.

$$\frac{5k}{5} = 1k = k$$

$$5k = 30$$

$$\frac{5k}{5} = \frac{30}{5}$$

$$k = 6$$

Divide to solve a multiplication equation.

Check: $5k = 30$

Replace the variable with the solution.

$$5(6) \stackrel{?}{=} 30$$

$$30 \stackrel{?}{=} 30 \quad ✔$$

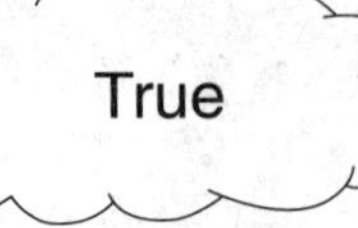

Solve and check.

5. $2w = 16$ **6.** $4b = 24$ **7.** $9z = 45$ **8.** $10m = 40$

_______________ _______________ _______________ _______________

Holt Mathematics

Challenge
Shape Up

Find the value of each shape in Exercises 1–8. Then use the values to answer the questions below.

1.

2 🦋 = 32

4 🐞 = 🦋

🦋 = __; 🐞 = __

2.

3 ⬤ = 18

2 🥁 = ⬤

⬤ = __; 🥁 = __

3.

4 ✳ = 8

2 🪁 = ✳

✳ = __; 🪁 = __

4.

12 ☆ = 108

3 ☾ = ☆

☆ = __; ☾ = __

5.

24 🐚 = 192

32 🐚 = 0

🐚 = __; 🐚 = __

6.

39 ❀ = 117

3 🍃 = ❀

❀ = __; 🍃 = __

7.

4 🍎 = 🍌 + 🍐

🍌 = 🍐 + 🍎

2 🍌 = 30

🍌 = __; 🍎 = __; 🍐 = __

8.

2 ⬭ = ◣

5 ⬡ = ⬭ + ◣

3 ⬭ = 15

⬭ = __; ◣ = __; ⬡ = __

9. What was the population of the United States in 1610?

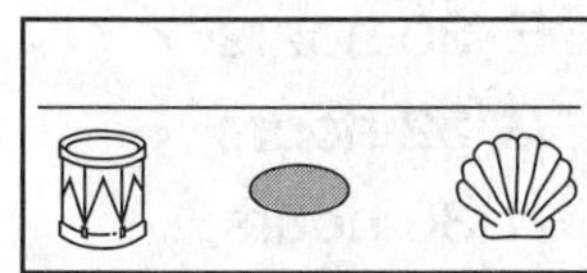

10. What was the population of the United States in 2000?

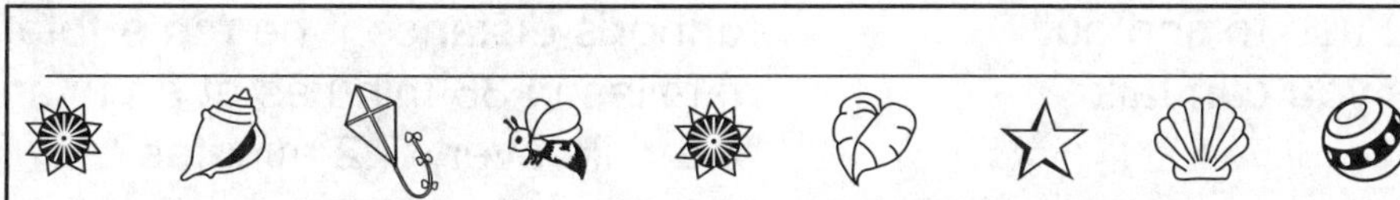

95

Holt Mathematics

Problem Solving
Solving Equations by Multiplying or Dividing

Write the correct answer.

1. The Panama Canal cost $387,000,000 to build. Each ship pays $34,000 to pass through the canal. How many ships had to pass through the canal to pay for the cost to build it?

2. The rate of exchange for currency changes daily. One day you could get $25 for 3,302.75 Japanese yen. Write and solve a multiplication equation to find the number of yen per dollar on that day.

3. Franklin D. Roosevelt was in office as president for 12 years. This is three times as long as Jimmy Carter was president. Write and solve an equation to show how long Jimmy Carter was president.

4. The mileage from Dallas to Miami is 1,332 miles. To the nearest hour, how many hours would it take to drive from Dallas to Miami at an average speed of 55 mi/h?

Choose the letter for the best answer.

5. The total bill for a bike rental for 8 hours was $38. How much per hour was the rental cost?

   A $8 per hour

   B $4.75 per hour

   C $30 per hour

   D $5.25 per hour

6. If a salesclerk earns $5.75 per hour, how many hours per week does she work to earn her weekly salary of $207?

   F 30 hours

   G 32 hours

   H 36 hours

   J 4 hours

7. At a cost of $0.07 per minute, which equation could you use to find out how many minutes you can talk for $3.15?

   A $0.07 \div m = $3.15

   B $3.15 \cdot m = $0.07

   C $0.07m = $3.15

   D $0.07 \div $3.15 = m

8. Which equation shows how to find a runner's distance if he ran a total of m miles in 36 minutes at an average of a mile every 7.2 minutes?

   F 36 \div m = 7.2

   G 7.2 \div m = 36

   H 36m = 7.2

   J 7.2 \div 36 = m

Holt Mathematics

LESSON 1-12 Reading Strategies
Follow a Procedure

The opposite of multiplication is division: $\longrightarrow$ $12 \cdot 3 = 36$, and $36 \div 3 = 12$
The opposite of division is multiplication: $\longrightarrow$ $48 \div 12 = 4$, and $4 \cdot 12 = 48$
From these examples you can see that:
division "undoes" multiplication, and **multiplication "undoes" division.**
To solve multiplication and division equations:

- Get the variable by itself on one side of the equation.

- Keep the equation in balance by using the same operation on both sides.

Example:

$84 = 7x$ $\longleftarrow$ Get the variable by itself. This is a multiplication equation, so divide to "undo" the multiplication.

$\dfrac{84}{7} = \dfrac{7x}{7}$ $\longleftarrow$ Rewrite the equation to show that both sides are divided by 7.

$12 = x$ $\longleftarrow$ This is the solution after dividing both sides by 7.

Check using 12 in place of x:
$84 \stackrel{?}{=} 7(12)$
$84 = 84$, so $x = 12$ is the solution.

Example:

$\dfrac{m}{15} = 8$ $\longleftarrow$ Get the variable by itself. Multiply to "undo" division.

$\dfrac{m}{15} \cdot 15 = 8 \cdot 15$ $\longleftarrow$ Rewrite the equation to show that both sides are multiplied by 15.

$m = 120$ $\longleftarrow$ This is the solution after multiplying both sides by 15.

Check by using 120 in place of m.
$\dfrac{120}{15} \stackrel{?}{=} 8$

$8 = 8$, so $m = 120$ is the solution.

Use $108 = 9y$ for Exercises 1–3.

1. What operation will you use to solve the equation?

____

2. Rewrite the equation using the inverse operation on both sides.

____

3. What is the value of y?

____

Holt Mathematics

Puzzles, Twisters & Teasers

LESSON 1-12 *Doctor! Doctor!*

Find the solution for each equation below. Write the letter of each variable on the line above the correct answer at the bottom of the page to solve the riddle.

1. $a \div 25 = 4$ _________

2. $w \times 18 = 18$ _________

3. $r \div 8 = 5$ _________

4. $3h = 96$ _________

5. $72 = 8d$ _________

6. $12 = y \div 4$ _________

7. $17 = n \div 8$ _________

8. $85 = 17o$ _________

9. $3e = 63$ _________

10. $9 = u \div 3$ _________

11. $6b = 222$ _________

12. $7m = 84$ _________

13. $150 = 3t$ _________

14. $9s = 99$ _________

Patient: Doctor! Doctor! I feel like an umbrella!

Doctor: _____ _____ _____,   _____ _____ _____
       1   32   48   48   5   27

    _____ _____ _____ _____   _____ _____
    12 27 11 50   37 21

    _____ _____ _____ _____ _____   _____ _____ _____
    27 136 9 21 40   50 32 21

    _____ _____ _____ _____ _____ _____ _____.
    1 21 100 50 32 21 40

Holt Mathematics

Write a rule that describes the pattern. Then write the next three numbers in the pattern.

1. 7, 10, 13, 16, __19__, __22__, __25__, ...
The pattern is
add 6 to each number to get the
next number

2. 81, 70, 59, 48, __37__, __26__, __15__, ...
The pattern is
subtract 11 from each number to
get the next number

3. 2, 4, 8, 16, __32__, __63__, __128__, ...
The pattern is
multiply each number by 2 to get
the next number

4. 1, 3, 6, 10, 15, __21__, __28__, __36__, ...
The pattern is
add 1 more than you did the
time before

Write a rule that describes the pattern. Then draw the next three figures in the pattern.

5. 
The pattern is _____ a circle and two triangles

6. 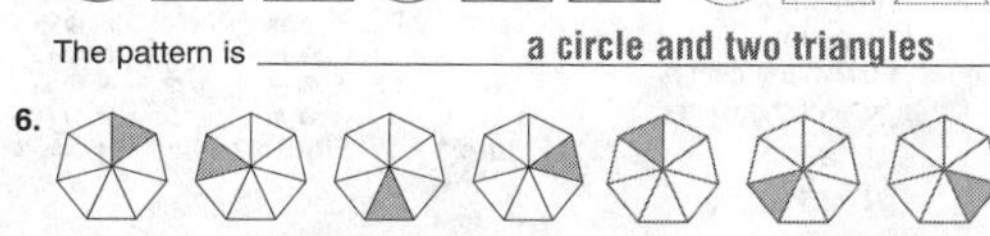
The pattern is is to shade every other triangle in a counterclockwise direction.

7. Complete the table so that it shows the number of dots in each figure.

Figure 1   Figure 2   Figure 3   Figure 4   Figure 5

Figure	1	2	3
Number of Dots	2	6	10

How many dots are in the fifth figure? __25__
Use drawings to justify your answer.

3   **Holt Mathematics**

Identify a possible pattern. Use the pattern to write the next three numbers.

1. 41, 37, 33, 29, __25__, __21__, __17__, ...
Subtract 4 from each number to
get the next number.

2. 50, 52, 56, 62, __70__, __80__, __92__, ...
Add 2 more than you did the
time before.

3. 320, 160, 80, 40, __20__, __10__, __5__, ...
Divide each number by 2 to get
the next number.

4. 24, 40, 56, 72, __88__, __104__, __120__, ...
Add 16 to each number to get the
next number.

Identify a possible pattern. Use the pattern to draw the next three figures.

5. 
_____ Alternate triangles and squares with circles between them.

6. 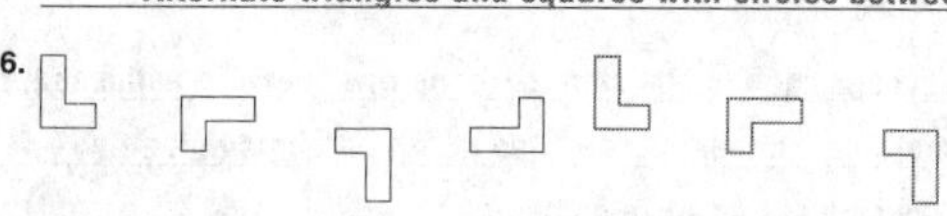
_____ Rotate the figure 90° clockwise each time.

7. Complete the table so that it shows the number of dots in each figure.

Figure 1   Figure 2   Figure 3   Figure 4   Figure 5

Figure	1	2	3
Number of Dots	1	4	9

How many dots are in the fifth figure of the pattern? __25__
Use drawings to justify your answer.

4   **Holt Mathematics**

Identify a possible pattern. Use the pattern to write the missing numbers.

1. 72, 95, 118, 141, __164__, __187__, __210__, ...
Add 23 to each number to get the
next number.

2. 65, __58__, 51, 44, __37__, __30__, 23, ...
Subtract 7 from each number to
get the next number.

3. 4, 12, __36__, 108, 324, 972, __2,916__, ...
Multiply each number by 3 to get
the next number.

4. __50__, __49__, 47, 44, 40, __35__, 29, ...
Subtract 1 more than you did the
time before.

Look for a possible pattern. Use the pattern to draw the next three figures.

5. 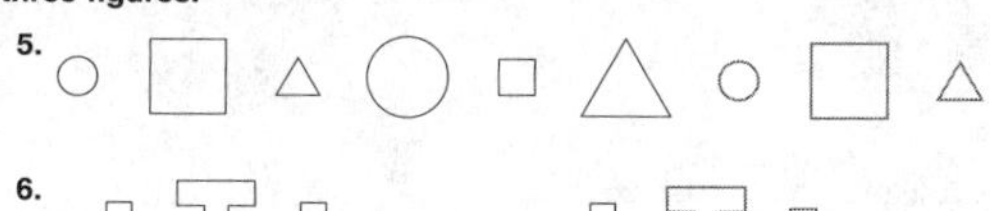

6. 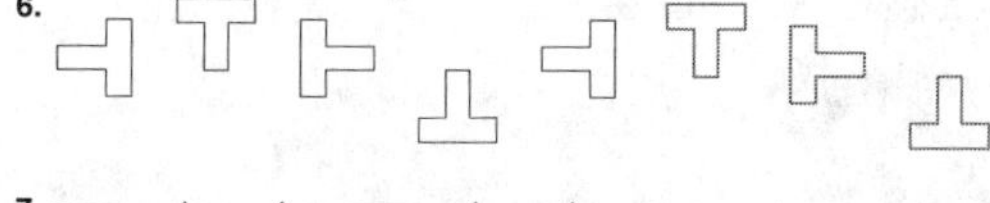

7.
| 7 | 15 | 23 | 31 | 39 | 47 | 55 | 63 | 71 | 79 |

8. Complete the table so that it shows the number of squares in each figure.

Figure 1   Figure 2   Figure 3   Figure 4   Figure 5

Figure	1	2	3
Number of Squares	2	3	12

How many squares are in the fifth figure of the pattern? __30__
Use drawings to justify your answer.

5   **Holt Mathematics**

To identify a number pattern, ask: What can I do to each number to get the number that comes next?

The pattern is to add 4 to get the next number. Use the pattern to get the next three numbers.

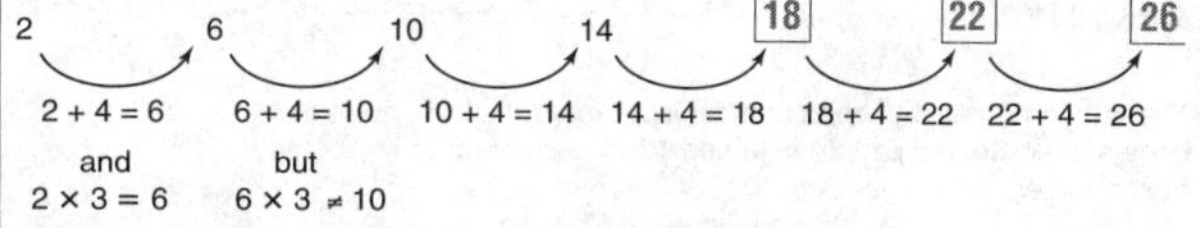

2 + 4 = 6   6 + 4 = 10   10 + 4 = 14   14 + 4 = 18   18 + 4 = 22   22 + 4 = 26
and   but
2 × 3 = 6   6 × 3 ≠ 10

Identify the pattern. Use the pattern to find the next three numbers.

1. 28, 25, 22, __19__, __16__, __13__, ...
Subtract __3__ to get the next number.
22 − __3__ = 19
__19__ − __3__ = 16
__16__ − __3__ = 13

2. 6, 24, 42, 60, __78__, __94__, __112__, ...
Add __18__ to get the next number.
60 + __18__ = 78
__78__ + __18__ = 94
__94__ + __18__ = 112

3. 4, 12, 20, 28, __36__, __44__, __52__, ...
Add 8 to get the next number.

4. 3, 6, 12, 24, __48__, __96__, __192__, ...
Multiply by 2 to get the next number.

To identify a geometric pattern, ask: How can I change each figure to get the next figure?

The pattern is to shade the next square in a clockwise direction. Use the pattern

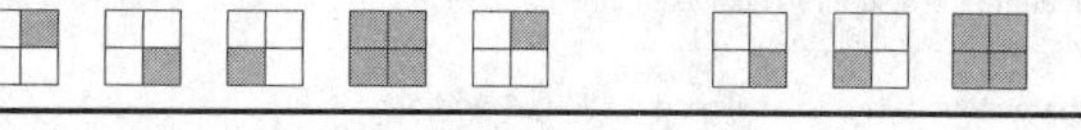

Draw the next three figures in each pattern.

5. 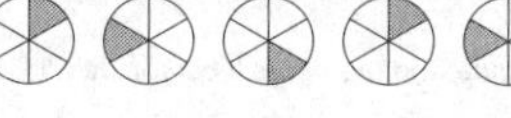 

6.  

6   **Holt Mathematics**

Challenge
Pascal's Triangle

The pattern at the right is called Pascal's Triangle. This pattern is named after the French mathematician, Blaise Pascal, because he wrote about its properties.

Use Pascal's Triangle to solve problems 1-5.

Row 0			1			
Row 1			1 1			
Row 2			1 2 1			
Row 3			1 3 3 1			
Row 4			1 4 6 4 1			
Row 5			1 5 10 10 5 1			
Row 6			? ? ? ? ? ? ?			
Row 7			? ? ? ? ? ? ?			
Row 8			? ? ? ? ? ? ?			
Row 9			? ? ? ? ? ? ?			

1. Write the sum of the numbers in each of rows 1-5. What pattern do you see?

   <u>2; 4; 8; 16; 32; the sum doubles for each row.</u>

2. Use the pattern that you found in problem 1 to predict the sum for row 9. <u>512</u>

3. Look at the numbers in row 4 and the numbers they are connected to in row 3. Then look at the numbers in row 5 and the numbers they are connected to in row 4. What pattern do you see?

   <u>If a number is connected to one number in the row above, it is the same as that number. If a number is connected to two numbers in the row above, it is the sum of those two numbers.</u>

4. Use the pattern you found in problem 3 to find the numbers for rows 6-9.

   row 6: <u>1, 6, 15, 20, 15, 6, 1</u>

   row 7: <u>1, 7, 21, 35, 35, 21, 7, 1</u>

   row 8: <u>1, 8, 28, 56, 70, 56, 28, 8, 1</u>

   row 9: <u>1, 9, 36, 84, 126, 126, 84, 36, 9, 1</u>

5. Find the sum of the numbers in row 9. How does it compare to your prediction in problem 2? <u>512; it matches.</u>

7

Holt Mathematics

Problem Solving
Numbers and Patterns

Write the correct answer.

1. Trains leave Peapack station at 6:40 A.M., 7:16 A.M., 7:32 A.M., and 7:48 A.M. Assume that the pattern continues. If you arrive at the station at 8:30 A.M., at what time will the next scheduled train leave?

   <u>8:36 A.M.</u>

2. Suppose the pattern 8, 16, 24, 32, ... is continued forever. Will the number 174 appear in the pattern? Why or why not?

   <u>No; the pattern is multiples of 8, and 174 is not a multiple of 8.</u>

3. A water tank holding 100 gallons begins to leak at 6:00 P.M. At 7:00 P.M. the tank has 94 gallons. At 8:00 P.M the tank has 88 gallons. At 9:00 P.M. the tank has 82 gallons. If the pattern continues, how many gallons will be left at 11:00 P.M.?

   <u>70 gallons</u>

4. Awilda is making a necklace with sphere, pyramid, and cube shaped beads. If she continues the pattern below, what are the shapes of the next five beads?

   <u>sphere, pyramid, sphere, sphere, cube</u>

Choose the letter for the best answer.

5. The drawing at the right shows the first three figures in a pattern. If the pattern continues, how many circles will be in the fifth figure of the pattern?

   A 20   C 27

   B 24   (D) 35

   Figure 1   Figure 2   Figure 3

6. The table shows the number of handshakes if each person in a group shakes each other's hand once. How many handshakes will there be if there are 9 people?

   F 20   (H) 36

   G 30   J 45

People	2	3	4	5	6
Number of Handshakes	1	3	6	10	15

7. Melissa wrote the following number pattern: 71, 66, 59, 50, 39, ... What is the next number in the pattern?

   (A) 26   C 30

   B 27   D 35

8. By July 1, Bob saved $120. By July 8, he saved $185. By July 15, Bob saved $250. If the pattern continues, how much will he have saved by July 29?

   F $315   H $390

8

Holt Mathematics

Reading Strategies
Identify Relationships

To identify and extend a **number pattern**, you must find the **relationship** between each number in the pattern and the number that it comes after.

4, 16, 28, 40, □, □, □, ...

4	16	28	40	52	64	76
	+12	+12	+12	+12	+12	+12

The relationship is that each number is 12 more than the number it comes after. So, the pattern is to add 12 to each number to get the next number.

To identify and extend a **geometric pattern**, you must find the **relationship** between each figure in the pattern and the figure that it comes after.

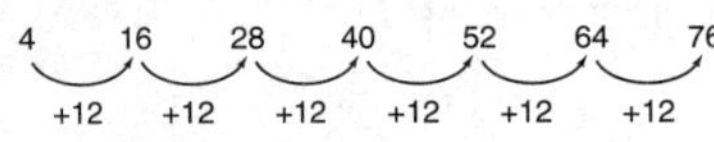

The relationship is that the shaded triangle moves one triangle counterclockwise from the figure it comes after.

So, the next three figures in the pattern are:

Use the pattern 64, 55, 46, 37, □, □, □, ... to answer questions 1 and 2.

1. What is the relationship between each number and the number it comes after?

   <u>Each number in the pattern is 9 less than the number it comes after.</u>

2. What are the next three numbers in the pattern?

   <u>28, 19, 10</u>

Use the pattern below to answer questions 3 and 4.

 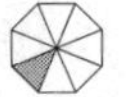 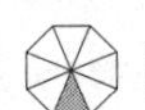  

3. What is the relationship between each figure and the figure it comes after?

   <u>The figure rotates 90° clockwise each time.</u>

9

Holt Mathematics

Puzzles, Twisters & Teasers
It's in the Cards!

When is it dangerous to play cards?

Write the next number or draw the next shape in each pattern. Write the letter on the line above the correct answer at the bottom of the page to solve the riddle.

T   75, 59, 43, 37, <u>21</u>, ...

E   7, 14, 28, 56, <u>112</u>, ...

H   12, 25, 39, 54, <u>70</u>, ...

J   243, 81, 27, 9, <u>3</u>, ...

W   28, 34, 42, 52, <u>64</u>, ...

K   114, 91, 68, 45, <u>22</u>, ...

R   2, 8, 32, 128, <u>512</u>, ...

N   70, 56, 41, 25, <u>8</u>, ...

S   5, 8, 14, 23, <u>35</u>, ...

O   6, 30, 54, 78, <u>102</u>, ...

L   6, 18, 54, 162, <u>486</u>, ...

I   

D

W	H	E	N		T	H	E
64	70	112	8		21	70	112

J	O	K	E	R		I	S		W	I	L	D
3	102	22	112	512		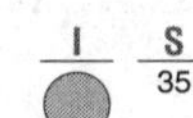	35		64		486	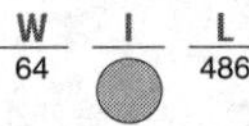

10

Holt Mathematics

Holt Mathematics

Practice A
Exponents

Multiply.

1. 4^2 	2. 2^3 	3. 6^2

$4 \cdot 4 = \underline{16}$ 	$2 \cdot 2 \cdot 2 = \underline{8}$ 	$6 \cdot 6 = \underline{36}$

4. 9^2 	5. 4^3 	6. 3^5 	7. 7^0

$\underline{81}$ 	$\underline{64}$ 	$\underline{243}$ 	$\underline{1}$

8. 10^2 	9. 3^4 	10. 9^1 	11. 2^5

$\underline{100}$ 	$\underline{81}$ 	$\underline{9}$ 	$\underline{32}$

Use an exponent and the given base to write each number.

12. 25, base 5 	13. 3, base 3 	14. 8, base 2 	15. 1, base 4

$\underline{5^2}$ 	$\underline{3^1}$ 	$\underline{2^3}$ 	$\underline{4^0}$

16. 81, base 9 	17. 64, base 4 	18. 64, base 8 	19. 9, base 3

$\underline{9^2}$ 	$\underline{4^3}$ 	$\underline{8^2}$ 	$\underline{3^2}$

20. 36, base 6 	21. 16, base 2 	22. 27, base 3 	23. 400, base 20

$\underline{6^2}$ 	$\underline{2^4}$ 	$\underline{3^3}$ 	$\underline{20^2}$

24. The first day, Jessie has \$2. The second day, she has twice as much money as the first day. The third day, she has twice as much money as the second day. Write the amount of money she has on the third day in exponential form. Then write the amount in standard form.

$\underline{2^3; \$8}$

25. Kevin runs 5 miles on Monday. The total number of miles he runs that week is 5 times the number of miles he runs on Monday. How many miles does Kevin run that week?

$\underline{25 \text{ miles}}$

  	11 	**Holt Mathematics**

Practice B
Exponents

Find each value.

1. 5^2 	2. 2^4 	3. 3^3 	4. 7^2

$\underline{25}$ 	$\underline{16}$ 	$\underline{27}$ 	$\underline{49}$

5. 4^4 	6. 12^2 	7. 10^3 	8. 11^1

$\underline{256}$ 	$\underline{144}$ 	$\underline{1,000}$ 	$\underline{11}$

9. 1^6 	10. 20^2 	11. 6^3 	12. 7^3

$\underline{1}$ 	$\underline{400}$ 	$\underline{216}$ 	$\underline{343}$

Write each number using an exponent and the given base.

13. 16, base 4 	14. 25, base 25 	15. 100, base 10 	16. 125, base 5

$\underline{4^2}$ 	$\underline{25^1}$ 	$\underline{10^2}$ 	$\underline{5^3}$

17. 32, base 2 	18. 243, base 3 	19. 900, base 30 	20. 121, base 11

$\underline{2^5}$ 	$\underline{3^5}$ 	$\underline{30^2}$ 	$\underline{11^2}$

21. 3,600, base 60 	22. 256, base 4 	23. 512, base 8 	24. 196, base 14

$\underline{60^2}$ 	$\underline{4^4}$ 	$\underline{8^3}$ 	$\underline{14^2}$

25. Damon has 4 times as many stamps as Julia. Julia has 4 times as many stamps as Claire. Claire has 4 stamps. Write the number of stamps Damon has in both exponential form and standard form.

$\underline{4^3 \text{ stamps; } 64 \text{ stamps}}$

26. Holly starts a jump rope exercise program. She jumps rope for 3 minutes the first week. In the second week, she triples the time she jumps. In the third week, she triples the time of the second week, and in the fourth week, she triples the time of the third week. How many minutes does she jump rope during the fourth week?

$\underline{81 \text{ minutes}}$

  	12 	**Holt Mathematics**

Practice C
Exponents

Find each value.

1. 3^8 	2. 7^5 	3. 50^2

$\underline{6,561}$ 	$\underline{16,807}$ 	$\underline{2,500}$

4. 9^4 	5. 30^3 	6. 12^5

$\underline{6,561}$ 	$\underline{27,000}$ 	$\underline{248,832}$

Compare. Write <, >, or =.

7. $7^2 \boxed{>} 48$ 	8. $9^3 \boxed{<} 810$ 	9. $6^4 \boxed{>} 10^3$

10. $100,000 \boxed{<} 10^6$ 	11. $2^7 \boxed{<} 6^3$ 	12. $4^6 \boxed{=} 8^4$

Write each number using an exponent and the given base.

13. 343, base 7 	14. 625, base 5 	15. 1,728, base 12

$\underline{7^3}$ 	$\underline{5^4}$ 	$\underline{12^3}$

16. 225, base 15 	17. 1,000,000, base 100 	18. 1,225, base 35

$\underline{15^2}$ 	$\underline{100^3}$ 	$\underline{35^2}$

19. 6,561, base 9 	20. 2,187, base 3 	21. 8,000, base 20

$\underline{9^4}$ 	$\underline{3^7}$ 	$\underline{20^3}$

22. Jacob has 7 times as many postcards as Austin has. Austin has 7 times as many postcards as Angela has. Angela has 7 times as many postcards as Samuel has. Samuel has 7 postcards. Write the number of postcards Jacob has in both exponential form and standard form.

$\underline{7^4 \text{ postcards; } 2,401 \text{ postcards}}$

  	13 	**Holt Mathematics**

Reteach
Exponents

The exponent tells you how many times to multiply the base by itself.

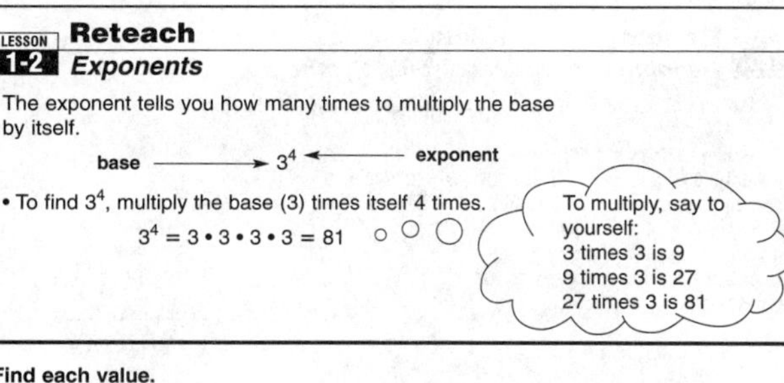

• To find 3^4, multiply the base (3) times itself 4 times.

$3^4 = 3 \cdot 3 \cdot 3 \cdot 3 = 81$

Find each value.

1. $4^3 = \underline{4} \cdot \underline{4} \cdot \underline{4} = 64$

2. $1^5 = \underline{1} \cdot \underline{1} \cdot \underline{1} \cdot \underline{1} \cdot \underline{1} = \underline{1}$

3. 5^2 	4. 2^3 	5. 3^3 	6. 6^2

$\underline{25}$ 	$\underline{8}$ 	$\underline{27}$ 	$\underline{36}$

7. 8^2 	8. 4^1 	9. 5^3 	10. 2^4

$\underline{64}$ 	$\underline{4}$ 	$\underline{125}$ 	$\underline{16}$

• You can write 64 using an exponent with the base 8.

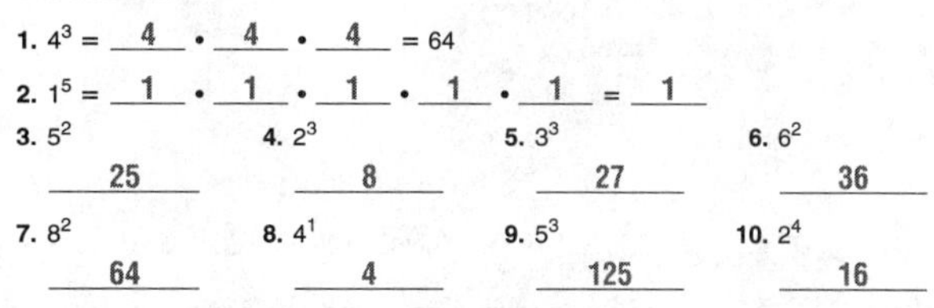

Write each number using an exponent and the given base.

11. 216, base 6: $= \underline{36 \cdot 6}$ and $36 \cdot 6 = \underline{6 \cdot 6 \cdot 6}$ so, $216 = \underline{6^3}$.

12. 16, base 4 	13. 8, base 2 	14. 9, base 3

$\underline{4^2}$ 	$\underline{2^3}$ 	$\underline{3^2}$

15. 81, base 9 	16. 27, base 3 	17. 49, base 7

$\underline{9^2}$ 	$\underline{3^3}$ 	$\underline{7^2}$

  	14 	**Holt Mathematics**

Challenge
Money Grows

If an investment of $50 doubles in value, your investment will be worth $50 • 2 = $100.

If it doubles again, your investment will be worth $50 • 2 • 2 = $200. You can also write this as $50 • 2^2 = $200.

If your money doubles a third time, it will be worth $50 • 2^3 = $400.

Write the correct answer.

1. If you invest $400, how much will it be worth if it doubles in value twice?

   $1,600

2. If an investment of $550 doubles in value twice, how much will it be worth?

   $2,200

3. If you invest $75, how much will it be worth if it doubles in value 4 times?

   $1,200

4. If you invest $125, how much will it be worth if it doubles in value 4 times?

   $2,000

5. If an investment of $75 triples in value 3 times, how much will it be worth?

   $2,025

6. If an investment of $125 triples in value 4 times, how much will it be worth?

   $10,125

When Jasmine turned 13, she started saving $5 per month until she turned 18. She then invested her total savings so that it doubled in value every 7 years.

7. How much money did Jasmine save by the time she turned 18? **$300**

8. How many times will her investment double from age 18 to age 60? **6 times**

9. How much will Jasmine's investment be worth when she turns 60? **$19,200**

When Jake turned 13, he started saving $1.50 per week until he turned 19. He then invested his total savings so that it tripled in value every 12 years.

10. How much money did Jake save by the time he turned 19? **$468**

11. How many times will his investment triple from age 19 to age 55? **3 times**

12. How much will Jake's investment be worth when he turns 55? **$12,636**

15

Holt Mathematics

Problem Solving
Exponents

Write the correct answer.

1. The cells of the bacteria *E. coli* can double every 20 minutes. If you begin with a single cell, how many cells can there be after 4 hours?

   4,096 cells

2. The population of metropolitan Orlando, Florida, has doubled about every 16 years since 1960. In 2000, the population was 1,644,561. At this doubling rate, what could the population be in 2048?

   13,156,488

3. A prizewinner can choose Prize A, $2,000 per year for 15 years, or Prize B, 3 cents the first year, with the amount tripling each year through the fifteenth year. Which prize is more valuable? How much is it worth?

   Prize B: $143,489.07

4. Maria had triplets. Each of her 3 children had triplets. If the pattern continued for 2 more generations, how many great-great-grandchildren would Maria have?

   81 great-great-grandchildren

Choose the letter for the best answer.

5. A theory states that the CPU clock speed in a computer doubles every 18 months. If the clock speed was 33 MHz in 1991, how can you use exponents to find out how fast the clock speed is after doubling 3 times?

   A 33^3
   B 3^2 • 33
   C 2^3 • 33
   D 33^2

6. The classroom is a square with a side length of 13 feet and an area of 169 square feet. How can you write the area in exponential form?

   F 2^{13} H 3^{13}
   G 13^2 J 13^3

7. In 2000, Wake County, North Carolina, had a population of 610,284. This is about twice the population in 1980. If the county grows at the same rate every 20 years, what will its population be in 2040?

   A 915,426   C 1,830,852
   B 1,220,568   D 2,441,136

8. The number of cells of a certain type of bacteria doubles every 45 minutes. If you begin with a single cell, how many cells could there be after 6 hours?

   F 64   H 360
   G 256   J 540

16

Holt Mathematics

Reading Strategies
Reading and Understanding Symbols

Exponents are an efficient way to express repeated multiplication.

3^5 → is read "3 to the fifth power."
3^5 means **3 is a factor 5 times:** $3 \times 3 \times 3 \times 3 \times 3$.
$3^5 = 243$ → is read "3 to the fifth power equals 243," or, "The value of 3 to the fifth power is 243."

The **base** names the factor. The **exponent** tells how many times to repeat the base as a factor.

base → 3^5 ← exponent

You can describe some numbers as the product of a repeated factor.

→ $64 = 4 \times 4 \times 4$

Then write the product with a base and an exponent.

→ $64 = 4^3$

Answer each question.

1. How do you read 5^3? **five to the third power**

2. What does 5^3 mean? **five is a factor 3 times: $5 \times 5 \times 5$**

3. What is the value of 5^3? **125**

4. Write about the difference between 5^3 and 3^5.

   Sample answer: 5^3 means $5 \times 5 \times 5 = 125$, and 3^5 means $3 \times 3 \times 3 \times 3 \times 3 = 243$.

5. Write 32 using 2 as the repeated factor.

   $32 = 2 \times 2 \times 2 \times 2 \times 2$

6. Write 32 using an exponent and a base of 2. **$32 = 2^5$**

7. Write 216 using 6 as the repeated factor.

   $6 \times 6 \times 6 = 216$

8. Write 216 using an exponent and a base of 6. **$216 = 6^3$**

17

Holt Mathematics

Puzzles, Twisters & Teasers
Explore Your Power Base!

Why was the computer tired when it got home?

To find out, solve each problem. Write the letter on the line above the correct answer at the bottom of the page. Each answer must match exactly. Some letters will not be used.

E $3^4 = $ **81**

T 49, base 7 = **7^2**

K $10^3 = $ **1,000**

B 81, base 9 = **9^2**

I $2^6 = $ **64**

D $5^2 = $ **25**

W $18^1 = $ **18**

C 25, base 5 = **5^2**

A 27, base 3 = **3^3**

H $14^0 = $ **1**

V $12^2 = $ **144**

R 256, base 4 = **4^4**

O 216, base 6 = **6^3**

U $5^3 = $ **125**

S $6^2 = $ **36**

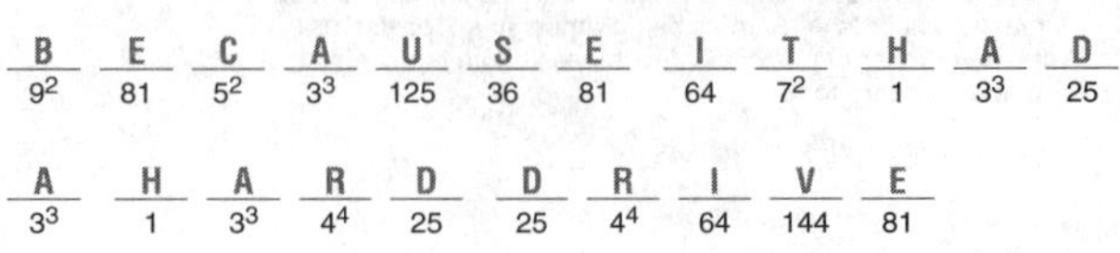

B	E	C	A	U	S	E	I	T	H	A	D
9^2	81	5^2	3^3	125	36	81	64	7^2	1	3^3	25

A	H	A	R	D	D	R	I	V	E
3^3	1	3^3	4^4	25	25	4^4	64	144	81

18

Holt Mathematics

Holt Mathematics

Choose the letter for the best answer.

1. The height of a flagpole
 A millimeters   Ⓒ meters
 B centimeters   D kilometers

2. The mass of a bicycle
 F milligrams   H grams
 Ⓖ kilograms   J centimeters

3. The capacity of a spoon
 A liters   C kiloliters
 Ⓑ milliliters   D milliliters

4. The thickness of a piece of cardboard
 Ⓕ millimeters   H kilometers
 G centimeters   J meters

5. The mass of a banana
 A kilograms   C liters
 Ⓑ grams   D milligrams

6. The capacity of a swimming pool
 F kilograms   H liters
 G milliliters   Ⓙ kiloliters

Choose the letter of the best answer.

7. 320 cm = ☐ m
 A 0.032   Ⓒ 3.2
 B 0.32   D 32

8. 2.5 kg = ☐ g
 F 0.0025   H 250
 G 0.25   Ⓙ 2,500

9. 5,600 mL = ☐ L
 A 5,600,000   C 56
 B 56,000   Ⓓ 5.6

10. 780 m = ☐ km
 F 780,000   Ⓗ 0.78
 G 78,000   J 0.078

11. 0.35 g = ☐ mg
 A 0.00035   Ⓒ 350
 B 0.0035   D 3,500

12. 4.3 mm = ☐ cm
 F 0.043   H 430
 Ⓖ 0.43   J 4,300

13. Benjamin is 165 cm tall. His sister is 1.8 m tall. Who is shorter? Explain your answer.

 Benjamin; 1.8 m = 180 cm; 165 cm < 180 cm

14. Jan has a bag of grapes that has a mass of 0.9 kg. Terri has a bag of grapes that has a mass of 830 g. Whose grapes have the greater mass? Explain your answer.

 Jan's; 0.9 kg = 900 g; 900 g > 830 g

Holt Mathematics

Choose the most appropriate metric unit for each measurement. Justify your answer.

1. The capacity of a paper cup

 Possible answer: Milliliters; a cup holds much less than a large water bottle, which may hold about 1 L.

2. The mass of a small poodle.

 Possible answer: Kilograms; a small poodle has a greater mass than a textbook, which may have a mass of about 1 kg.

3. The width of a computer screen

 Possible answer: Centimeters; the width of a computer screen is less than the width of a doorway, which is about 1 m.

4. The mass of a pencil

 Possible answer: Grams; the mass of a pencil is similar to the mass of 10 paperclips, each of which may have a mass of about 1 g.

Convert each measure.

5. 496 mm to centimeters
 49.6 cm

6. 0.68 kg to grams
 680 g

7. 3,800 mL to liters
 3.8 L

8. 832 mg to grams
 0.832 g

9. 76 km to meters
 76,000 m

10. 2.9 cm to meters
 0.029 m

11. 0.041kL to liters
 41 L

12. 14.9 g to milligrams
 14,900 mg

13. 7,800 cm to meters
 78 m

14. Sam's laptop computer has a mass of 4.2 kg. Fred's laptop computer has a mass of 4,940 grams. Which computer has the lesser mass? Explain your answer.

 Sam's computer; 4.2 kg = 4,200 g; 4,200 g < 4,940 g

15. Elise makes a poster that is 1.5 m tall. Meg makes a poster that is 96 cm tall. Who makes a taller poster? Explain your answer.

 Elise; 1.5 m = 150 cm; 150 cm > 96 cm

Holt Mathematics

Choose the most appropriate metric unit for each measurement. Justify your answer.

1. The thickness of a piece of spaghetti

 Possible answer: Millimeters; the thickness of a piece of spaghetti is similar to the thickness of a dime, which is about 1 cm.

2. The capacity of a small sink

 Possible answer: liters; The capacity of a small sink might be similar to the capacity of about 5 large water bottles, each of which holds about 1 L.

3. The mass of an envelope

 Possible answer: Grams; the mass of an envelope is similar to the mass of a group of 10 paperclips, each of which may have a mass of about 1 g.

4. The capacity of a gas truck

 Possible answer: Kiloliters; the capacity of a gas truck is greater than the capacity of 2 large refrigerators, which may be about 1 kL.

Convert each measure.

5. 3.06 kL to liters
 3,060 L

6. 75 mm to centimeters
 7.5 cm

7. 604 g to kilograms
 0.604 kg

8. 0.95 m to centimeters
 95 cm

9. 5.6 L to milliliters
 5,600 mL

10. 48 mg to grams
 0.048 g

Compare. Write >, <, or = for ☐.

11. 0.5 cm = 5 mm

12. 4.6 kg > 462 g

13. 0.03 kL < 200 L

14. 53 m > 5,400 mm

15. 8,840 mL < 9 L

16. 416 mg > 0.4 g

17. Eleanor ran 3.5 km on Friday and 2.8 km on Saturday. Monroe ran 6,400 m on the same two days. Who runs a greater distance? Explain.
 Monroe. 3.5 km = 3,500 m; 2.8 km = 2,800 m; 3,500 m + 2,800 m = 6,300 m; 6,300 m < 6,400 m.

18. Tunnel A is 1.6 km long. Tunnel B is 740 meters long. Which tunnel is longer? How much longer?

 Tunnel A; 860 m

Holt Mathematics

You can use the relationships in this table to help you convert metric units.

Length	Mass	Capacity
1 cm = 10 mm	1 g = 1,000 mg	1 L = 1,000 mL
1 m = 100 cm = 1,000 mm	1 kg = 1,000 g	1 kL = 1,000 L
1 km = 100 m		

To change larger units to smaller units, multiply.

 5.3 L = ☐ mL
 5.3 × 1,000 = 5,300
 5.3 L = 5,300 mL

1 L = 1,000 mL
Liters are larger than milliliters, so multiply by 1,000.
Move the decimal point 3 places to the right: 5.300

To change smaller units to larger units, divide.

 46 cm = ☐ m
 46 ÷ 100 = 0.46
 46 cm = 0.46 m

1 m = 100 cm
Centimeters are smaller than meters, so divide by 100.
Move the decimal point 2 places to the left: 046

Convert each measure.

1. 15 cm to millimeters
 Think: 1 cm = **10** mm
 Multiply 15 by **10**.
 15 cm = **150** mm

2. 964 g to kilograms
 Think 1 kg = **1,000** g
 Divide 964 by **1,000**
 964 g = **0.964** kg

3. 4.9 kL to liters
 Multiply 4.9 by **1,000**
 4.9 kL = **4,900** mm

4. 3,531 cm to meters
 Divide 3,531 by **100**
 3,531 cm = **35.31** m

5. 468 mg to grams
 468 mg = **0.468** g

6. 0.69 kL to liters
 0.69 kL = **690** L

Holt Mathematics

Holt Mathematics

Challenge
Have a Ball!

The most commonly used metric units of mass are **milligrams (mg)**, **grams (g)**, and **kilograms (kg)**. However, there are other metric units of mass also. These units include **centigrams (cg)**, **decigrams (dg)**, **decagrams (dag)**, and **hectograms (hg)**. The table below shows the equivalent number of grams for each of these units.

Unit	1 kg	1 hg	1 dag	1 g	1 dg	1 cg	1 mg
Number of Grams	1,000 g	100 g	10 g	1 g	0.1 g	0.01 g	0.001 g

The table below shows the masses of balls that are used in different sports. Rewrite the table so that it shows the balls and their masses in order from greatest mass to least mass. Use the same unit to show each mass. **Masses may vary depending upon unit chosen.**

Ball	Mass
Baseball	14.88 dag
Basketball	0.65 kg
Football	4,252 dg
Paddleball	6,502 cg
Ping pong ball	2,450 mg
Softball	198.4 g
Soccer ball	4.536 hg
Tennis Ball	750 dg
Volleyball	28,000 cg

Ball	Mass
Basketball	650 g
Soccer ball	453.6 g
Footballl	425.2 g
Volleyball	280 g
Softball	198.4 g
Baseball	148.8 g
Tennis ball	75 g
Paddleball	65.02 g
Ping pong ball	2.45 g

23
Holt Mathematics

Problem Solving
Metric Measurements

Write the correct answer.

1. A porcupine has a mass of 27 kg. A cat has a mass of 6,300 g. Which animal has the greater mass?

   __Porcupine__

2. A faucet drips at the rate of 50 L per day. At this rate, how many days will it take for the faucet to drip a total of 1 kL?

   __20 days__

3. The distance from Midwood Library to Midwood Middle School is 0.845 km. The distance from Midwood High School to Midwood Library is 872 m. Which school is closer to the library? How much closer?

   __Midwood Middle School; 27 m__

4. A mouse has a mass of 18 g. A guinea pig has a mass of 0.85 kg. Is the mass of the two animals together greater or less than 1 kg? Explain your answer.

   __Less; 18 g = 0.018 kg; 0.018 kg +__
   __0.85 kg = 0.868 kg; 0.868 kg < 1 kg__

Choose the letter for the best answer.

The table shows the lengths and masses of some members of the cat family.

Members of the Cat Family		
Name	Length	Mass
Bobcat	864 mm	18,200 g
Jaguar	1.52 m	55 kg
Canada Lynx	91.4 cm	15,900 g
Mountain Lion	1.35 m	85 kg

5. Which cat has the greatest length?
   A bobcat
   (B) jaguar
   C Canada lynx
   D mountain lion

6. Which cat has the least mass?
   F bobcat
   G jaguar
   (H) Canada lynx
   J mountain lion

7. What is the difference in mass between the cat with the greatest mass and the cat with the least mass?
   A 66.8 kg
   (B) 69.1 kg
   C 7,400 g
   D 15,815 g

8. For which two cats is the difference in length closest to 0.5 m?
   (F) bobcat and mountain lion
   G bobcat and Canada lynx
   H jaguar and mountain lion
   J jaguar and Canada lynx

24
Holt Mathematics

Reading Strategies
Follow a Procedure

To convert metric units, follow these steps.

$653 \text{ mg} = \boxed{} \text{ g}$

Step 1: Determine if you are converting to a larger unit or a smaller unit.

1 mg < 1 g
You are converting to a larger unit.

Step 2: Choose the operation that you need to use to convert.
  • To convert to a smaller unit, you need to multiply.
  • To convert to a larger unit, you need to divide.

Since you are converting to a larger unit, you need to divide.

Step 3: Identify the number by which you must multiply or divide. That number is the number of smaller units that equal 1 larger unit.

1,000 mg = 1 g
So, you need to divide 653 mg by 1,000 to get the number of grams.

Step 4: Calculate the answer.

653 ÷ 1,000 = 0.653
653 mg = 0.653g

Answer these questions for $3.9 \text{ m} = \boxed{} \text{ cm.}$

1. Are you converting to a larger unit or a smaller unit? __a smaller unit__

2. What operation will you use to convert? __multiplication__

3. By what number do you need to multiply or divide? __100__

4. What is the answer to the problem? __390 cm__

Answer these questions for $14.8 \text{ L} = \boxed{} \text{ kL.}$

5. Are you converting to a larger unit or a smaller unit? __a larger unit__

6. What operation will you use to convert? __division__

7. By what number do you need to multiply or divide? __1,000__

8. What is the answer to the problem? __0.0148 kL__

25
Holt Mathematics

Puzzles, Twisters, & Teasers
Measure for Measure

Choose the most appropriate metric unit for each measurement. Circle the letter above your answer.

1. The capacity of a juice can
   (Y)   T
   milliliters   kiloliters

2. the length of a swimming pool
   H   (O)
   millimeters   meters

3. The mass of a bowling ball
   E   (U)
   milligrams   kilograms

4. The capacity of a walk-in freezer
   (R)   L
   kiloliters   milliliters

5. The width of a postcard
   (E)   I
   centimeters   kilometers

6. The mass of a button
   T   (A)
   kilograms   grams

Choose the measurement that is equivalent to the one given. Circle the letter above your answer.

7. 680 mg = ?
   T   (R)
   0.068 g   0.68 g

8. 13.7 cm = ?
   (D)   L
   137 mm   13.7 m

9. 0.49 kg = ?
   E   (R)
   4,900 g   490 g

10. 27.1 L = ?
   (U)   T
   0.0271 kL   271 mL

11. 964 mm = ?
   (M)   O
   0.964 m   9.64 cm

12. 0.78 m = ?
   T   (S)
   78 mm   78 cm

Start with problem number 1, and use the circled letters to solve the riddle.

What part of your body has the most rhythm?

Y   O   U   R

E   A   R   D   R   U   M   S

26
Holt Mathematics

Practice A
Powers of Ten and Scientific Notation

Choose the letter for the best answer.

1. $4 \cdot 10^2$
   A 4
   B 40
   C 400 *(circled)*
   D 100

2. $16 \cdot 10^0$
   F 1600
   G 16 *(circled)*
   H 10
   J 0

3. $9 \cdot 10^3$
   A 9,000 *(circled)*
   B 900
   C 90
   D 9

4. $17 \cdot 10^1$
   F 1.7
   G 17
   H 170 *(circled)*
   J 1700

Multiply.

5. $23 \cdot 10^2$
   2,300

6. $15 \cdot 10^4$
   150,000

7. $30 \cdot 10^2$
   3,000

8. $28 \cdot 10^3$
   28,000

9. $132 \cdot 10^2$
   13,200

10. $201 \cdot 10^3$
   201,000

11. $456 \cdot 10^2$
   45,600

12. $108 \cdot 10^4$
   1,080,000

Write the number in scientific notation.

13. 56,000
   5.6×10^4

14. 306,000
   3.06×10^5

15. 8,000,000
   8.0×10^6

16. 7,200,000
   7.2×10^6

17. 14,000,000
   1.4×10^7

18. 41.00
   4.1×10^1

19. 2,144,000
   2.144×10^6

20. $20.3 \cdot 10^5$
   2.03×10^6

21. Lake Huron covers an area of about 23,000 square miles. Write this number in scientific notation.

   2.3×10^4

22. The planet Mercury is about 3.6×10^7 miles from the sun. Write this number in standard form.

   36,000,000

Holt Mathematics

Practice B
Powers of Ten and Scientific Notation

Multiply.

1. $6 \cdot 10^3$
   6,000

2. $22 \cdot 10^1$
   220

3. $8 \cdot 10^2$
   800

4. $18 \cdot 10^0$
   18

5. $70 \cdot 10^2$
   7,000

6. $25 \cdot 10^3$
   25,000

7. $3 \cdot 10^4$
   30,000

8. $180 \cdot 10^3$
   180,000

9. $84 \cdot 10^4$
   840,000

10. $315 \cdot 10^2$
   31,500

11. $210 \cdot 10^3$
   210,000

12. $1,004 \cdot 10^3$
   1,004,000

13. $1,764 \cdot 10^1$
   17,640

14. $856 \cdot 10^0$
   856

15. $4,055 \cdot 10^3$
   4,055,000

16. $716 \cdot 10^4$
   7,160,000

Write each number in scientific notation.

17. 34,000
   3.4×10^4

18. 7,700
   7.7×10^3

19. 2,100,000
   2.1×10^6

20. 404,000
   4.04×10^5

21. 21,000,000
   2.1×10^7

22. 612.00
   6.12×10^2

23. 3,001,000
   3.001×10^6

24. $62.13 \cdot 10^4$
   6.213×10^5

25. Lake Superior covers an area of about 31,700 square miles. Write this number in scientific notation.

   3.17×10^4

26. Mars is about $1.42 \cdot 10^8$ miles from the sun. Write this number in standard form.

   142,000,000

27. In 2005, the population of China was about $1.306 \cdot 10^9$. What was the population of China written in standard form?

   1,306,000,000

28. A scientist estimates there are 4,800,000 bacteria in a test tube. How does she record the number using scientific notation?

   4.8×10^6

Holt Mathematics

Practice C
Powers of Ten and Scientific Notation

Multiply.

1. $5 \cdot 10^3$
   5,000

2. $471 \cdot 10^2$
   47,100

3. $39.5 \cdot 10^1$
   395

4. $200 \cdot 10^5$
   20,000,000

5. $7,025 \cdot 10^0$
   7,025

6. $5.7 \cdot 10^6$
   5,700,000

7. $66.25 \cdot 10^4$
   662,500

8. $9.01 \cdot 10^9$
   9,010,000,000

Write each number in scientific notation.

9. 25,000
   2.5×10^4

10. 9,900
   9.9×10^3

11. 9,700,000
   9.7×10^6

12. $95.6 \cdot 10^8$
   9.56×10^9

13. 23,000,000,000
   2.3×10^{10}

14. 110.00
   1.1×10^2

15. $301.9 \cdot 10^5$
   3.019×10^7

16. $73.55 \cdot 10^4$
   7.355×10^5

Write the missing number or numbers.

17. $1.23 \times 10^7 = 12,300$
   4

18. $8.3 \times 10^5 = ?$
   830,000

19. $112,000,000 = ? \times 10^8$
   1.12

20. $410,000 = ? \times 10^5$
   4.1

21. $7.7 \times 10^7 = ?$
   77,000,000

22. $2,950,000 = 2.95 \times 10^?$
   6

23. The Caspian Sea covers an area of about 143,250 square miles. Write this number in scientific notation.

   1.4325×10^5

24. The distance between Jupiter and Saturn is about 4.03×10^8 miles. Write this number in standard form.

   403,000,000

25. In 2002, there were about 3.005×10^7 pet dogs in Brazil and about 9.65×10^6 pet dogs in Japan. In which country were there more pet dogs?

   Brazil

26. A scientist estimates there are 37,000,000 bacteria in a petri dish. He records the number using scientific notation. What does he write?

   3.7×10^7

Holt Mathematics

Reteach
Powers of Ten and Scientific Notation

To multiply by a power of 10, use the exponent to find the number of zeros in the product.
• Multiply $42 \cdot 10^4$.

   $42 \cdot 10^4 = 420,000$

The exponent 4 tells you to write 4 zeros after 42 in the product.

Find each product.

1. $84 \cdot 10^3$
   The product should have __3__ zeros.
   $84 \cdot 10^3 = 84,000$

2. $61 \cdot 10^5$
   The product should have __5__ zeros.
   $61 \cdot 10^5 = 6,100,000$

3. $22 \cdot 10^6$
   22,000,000

4. $753 \cdot 10^3$
   753,000

5. $825 \cdot 10^2$
   82,500

6. $123 \cdot 10^1$
   1,230

• Write 926,000 in scientific notation.

First, write the digits before the zeros as a number greater than or equal to 1 and less than 10. The number must have only 1 digit to the left of the decimal point. That digit cannot be zero.

   0 1 2 3 4 5 6 7 8 9 10

Think: 9.26 is greater than 1 and less than 10.

Then multiply 9.26 by the power of 10 that gives 926,000 as the product.

   $9.2\,6\,0\,0\,0 = 9.26 \times 10^5$

The decimal point moves 5 places so the exponent is 5.

Write each number in scientific notation.

7. 5,100
   The decimal point moves __3__ places.
   $5,100 = 5\,.\,1 \times 10^3$

8. 1,840,000
   The decimal point moves __6__ places.
   $1,840,000 = 1\,.\,84 \times 10^6$

9. 641,000
   6.41×10^5

10. 47,300
   4.73×10^4

11. 8,250,000
   8.25×10^6

12. 703,000
   7.03×10^5

Holt Mathematics

Holt Mathematics

Challenge
Computer Bytes

Each byte in a computer's memory represents about one character. The major units of computer memory are kilobytes (KB), megabytes (MB), and gigabytes (GB).

1 kilobyte	= 1,000 bytes	1 KB	= 1,000 bytes
1 megabyte	= 1,000 kilobytes	1 MB	= 1,000 KB
1 gigabyte	= 1,000 megabytes	1 GB	= 1,000 MB

Write your answers using scientific notation.

1. In 1984, many personal computers had 64 KB of active (RAM) memory. How many bytes does this represent?

6.4×10^4 bytes

2. In 1992, many personal computers had 40 MB of hard drive memory. How many bytes does this represent?

4×10^7 bytes

3. In 1997, many personal computers had 1 GB of hard drive memory. How many bytes does this represent?

1×10^9 bytes

4. By 2005, many personal computers had 250 GB of hard drive memory. How many bytes does this represent?

2.5×10^{11} bytes

Ming saved his computer files on floppy disks. Each disk holds up to 1.44 MB of memory. He used these disks to transfer his files to another computer.

5. How many bytes could each floppy disk hold?

1.44×10^6 bytes

6. Ming's new computer has 120 GB of memory. How many disks could he transfer if each disk held 1.2 MB?

100,000 disks

Rachel decided to back up her hard drive's computer files by copying them onto compact disks (CDs). Each CD can hold up to 650 MB of memory, but Rachel saves only 600 MB on each.

7. How many bytes could each CD potentially hold?

6.5×10^8 bytes

8. If Rachel backs up 6 GB of memory, how many bytes of memory will she need?

6×10^9 bytes

31   **Holt Mathematics**

Problem Solving
Powers of Ten and Scientific Notation

Write the correct answer.

1. Earth is about 150,000,000 kilometers from the sun. Write this distance in scientific notation.

1.5×10^8 km

2. The planet Neptune is about 4.5×10^9 kilometers from the sun. Write this distance in standard form.

4,500,000,000 km

3. At the end of 2004, the U.S. federal debt was about $7 trillion, 600 billion. Write the amount of the debt in standard form and in scientific notation.

$7,600,000,000,000;$

7.6×10^{12}

4. Canada is about 1.0×10^7 square kilometers in size. Brazil is about 8,500,000 square kilometers in size. Which country has a greater area?

Canada

Choose the letter for the best answer.

5. China's population in 2001 was approximately 1,273,000,000. Mexico's population for the same year was about 1.02×10^8. How much greater was China's population than Mexico's?

A 1,375,000,000
B 1,274,020,000
Ⓒ 1,171,000,000
D 102,000,000

6. In mid-2001, the world population was approximately 6.137×10^9. By 2050, the population is projected to be 9.036×10^9. By how much will world population increase?

F 151,730,000
G 289,900,000
H 1,517,300,000
Ⓙ 2,899,000,000

7. The Alpha Centauri star system is about 4.3 light-years from Earth. One light-year, the distance light travels in 1 year, is about 6 trillion miles. About how many miles away from Earth is Alpha Centauri?

Ⓐ 2.58×10^{13} miles
B 6×10^{13} miles
C 1.03×10^{12} miles
D 2.58×10^9 miles

8. In the fall of 2001, students in Columbia, South Carolina, raised $440,000 to buy a new fire truck for New York City. If the money had been collected in pennies, how many pennies would that have been?

F 4.4×10^6
G 4.4×10^5
Ⓗ 4.4×10^7
J $4.4 \ 3 \ 10^8$

32   **Holt Mathematics**

Reading Strategies
Follow a Procedure

When you have an exponent with a base of 10, the number is called a **power of 10.**
You can use some simple rules to find products with powers of 10.

13×10^3
↓
$13 \times (10 \times 10 \times 10)$
↓
$13 \times 1,000$
↓
3.0̣0̣0̣,

Move the decimal point 3 places to the right. You need to add 3 zeros.

32.5×10^3
↓
$32.5 \times (10 \times 10 \times 10)$
↓
$32.5 \times 1,000$
↓
32,500.

Move the decimal point 3 places to the right. You need to add 2 zeros.

You can use powers of 10 to write large numbers in **scientific notation.** Scientific notation is used as a shortcut to write very large or very small numbers.

To write 268,000,000 in scientific notation:

Step 1: Move the decimal point to create a number between 1 and 10. 2.6̣8̣.0̣.0̣.0̣.0̣.0̣.(← Move the decimal point 8 places.

Step 2: The number of places the decimal point is moved is the value of the exponent. 2.68×10^8 ← So, the exponent is 8.

268,000,000 written in scientific notation is 2.68×10^8.

Use 2.8×10^5 to answer Exercises 1–4.

1. How many times is 10 a factor? _____ 5 times

2. Rewrite the number with 10 as a repeated factor.

$2.8 \times 10 \times 10 \times 10 \times 10 \times 10$

3. How many places will you move the decimal point? How many zeros will be in the product? _____ 5 places; 4

4. What is the product of 2.8×10^5? _____ 280,000

33   **Holt Mathematics**

Puzzles, Twisters & Teasers
Oh, the Power of Tens!

Substitute the correct number for the letter or letters in each equation. Use your answers to solve the riddle.

1. $24,500 = 2.45 \times 10^E$ _____ 4

2. $280,000 = 2.8 \times 10^P$ _____ 5

3. $592,000 = I \times 10^5$ _____ 5.92

4. $16,800 = C \times 10^4$ _____ 1.68

5. $5.4 \times 10^H = 540,000,000$ _____ 8

6. $24,400,000 = S \times 10^A$ _____ 7

What's a Martian's favorite snack?

S	P	A	C	E	C	H	I	P	S
2.44	5	7	1.68	4	1.68	8	5.92	5	2.44

34   **Holt Mathematics**

106

Holt Mathematics

Practice A
Order of Operations

Choose the letter for the best answer.

1. $75 + 12 \cdot 2$
   A 87   C 108
   (B) 99   D 174

2. $100 - 25 \div 5$
   F 15   H 80
   G 75   **(J) 95**

3. $50 - 18 \div 6 + 2$
   (A) 49   C 10
   B 40   D 4

4. $72 - 4^2 \cdot 2$
   F 32   H 56
   (G) 40   J 64

5. $(8 + 22) \div 5 + 5$
   A 30   **(C) 11**
   B 17.4   D 3

6. $3^3 - (9 \cdot 2 + 1)$
   F 19   **(H) 8**
   G 10   J −10

Simplify each expression.

7. $2^4 \div 8 + 5$

   _7_

8. $18 + 2(1 + 3^2)$

   _38_

9. $(16 \div 4) + 4 \cdot (2^2 - 2)$

   _12_

10. $2^3 - (3 \cdot 5 - 8)$

   _1_

11. $35 + 4^2 - (6 - 3)$

   _48_

12. $6 \cdot 7 - 3(4 + 1)$

   _27_

13. $100 \div 5 \cdot 2^2$

   _80_

14. $(5 + 2)^2 \div 7 - 6$

   _1_

15. $15 - 3 \cdot 4 \div 2 + 5$

   _14_

16. Leon rents a video game for $5. He returns the video game 3 days late. The late fee is $1 for each day late. Simplify the expression $5 + 3 \cdot 1$ to find out how much it costs Leon to rent the video game.

   $8

17. Olivia shovels snow for 6 neighbors. Each neighbor pays her $8. Two neighbors each give her a $2 tip as well. Simplify the expression $6 \cdot 8 + 2 \cdot 2$ to find out how much money Olivia earns in all.

   $52

35

Practice B
Order of Operations

Simplify each expression.

1. $15 \cdot 3 + 12 \cdot 2$
   _69_

2. $212 + 21 \div 3$
   _219_

3. $9 \cdot 3 - 18 \div 3$
   _21_

4. $65 - 36 \div 3$
   _53_

5. $100 - 9^2 + 2$
   _21_

6. $3 \cdot 5 - 45 \div 3^2$
   _10_

7. $54 \div 6 + 4 \cdot 6$
   _33_

8. $(6 + 5) \cdot 16 \div 2$
   _88_

9. $60 - 8 \cdot 12 \div 3$
   _28_

10. $45 - 3^2 \cdot 5$
   _0_

11. $52 - (8 \cdot 2 \div 4) + 3^2$
   _57_

12. $(2^3 + 10 \div 2) \cdot 3$
   _39_

13. $25 + 7(18 - 4^2)$
   _39_

14. $(6 \cdot 3 - 12)^2 \div 9 + 7$
   _11_

15. $4^3 - (3 + 12 \cdot 2 - 9)$
   _46_

16. $2^4 \div 8 + 5$
   _7_

17. $(1 + 2)^2 \cdot (3 - 1)^2 \div 2$
   _18_

18. $(16 \div 4) + 4 \cdot (2^2 - 2)$
   _12_

19. $2^5 - (3 \cdot 7 - 7)$
   _18_

20. $75 + 5^2 - (8 - 3)$
   _95_

21. $9 \cdot 6 - 5(10 - 3)$
   _19_

22. $96 \div 4 + 5 \cdot 2^2$
   _44_

23. $(15 - 6)^2 \div 3 - 3^3$
   _0_

24. $19 - 8 \cdot 5 \div 10 + 6 \div 3$
   _17_

25. Jared has $32. He buys 5 packs of trading cards that cost $3 each and a display book that costs $7. Simplify the expression $32 - (5 \cdot 3 + 7)$ to find out how much money Jared has left.

   $10

26. David buys 3 movie tickets for $6 each and 2 bags of popcorn for $2 each. Simplify the expression $3 \cdot 6 + 2 \cdot 2$ to find out how much money David spent in all.

   $22

36

Practice C
Order of Operations

Simplify each expression.

1. $25 \cdot 3 + 60 \cdot 2$
   _195_

2. $350 \div 5 + 12 \cdot 7$
   _154_

3. $3 \cdot 9 + 96 \div 4$
   _51_

4. $77 - 42 \div 7^1$
   _71_

5. $532 - 2^5 \div 4$
   _524_

6. $3(20 - 4^2) + 7$
   _19_

7. $270 \div 6 + 6^2$
   _81_

8. $(5 + 6)^2 + 18 \div 2$
   _130_

9. $10^2 - 25 \cdot 3 \div 5$
   _85_

10. $65 - 4^3 \cdot 1^7$
   _1_

11. $40 - (5 \cdot 2) + 8$
   _38_

12. $(6^2 + 4) \div 5$
   _8_

13. $2^4 \div 8 + 5$
   _7_

14. $(1 + 2)^2 \cdot (3 - 1)^2 \div 2$
   _18_

15. $(16 \div 4) + 4 \cdot (2^2 - 2)$
   _12_

Insert grouping symbols to make each statement true.

16. $18 + 2 \cdot 1 + 3^2 = 38$
   $18 + 2(1 + 3^2)$

17. $4 \cdot 2 - 2^2 \div 9 + 2 = 6$
   $(4 \cdot 2 - 2)^2 \div 9 + 2$

18. $3^3 - 9 \cdot 2 + 1 = 8$
   $3^3 - (9 \cdot 2 + 1)$

19. $2^3 - 3 \cdot 5 - 8 = 1$
   $2^3 - (3 \cdot 5 - 8)$

20. $35 + 4^2 - 6 - 3 = 48$
   $35 + 4^2 - (6 - 3)$

21. $6 \cdot 7 - 3 \cdot 4 + 1 = 27$
   $6 \cdot 7 - 3 \cdot (4 + 1)$

22. A group of students charges $7 to clean the exterior and $6 to clean the interior of a car. They clean 9 exteriors and 5 interiors. Simplify the expression $7 \cdot 9 + 6 \cdot 5$ to find out how much money the students raised in all.

   $93

23. Ariel has $65. She buys 5 books that cost $8.00 each, a bookmark that costs $2.00, and a magazine that costs $4.00. Simplify the expression $65 - (5 \cdot 8 + 2 + 4)$ to find out how much money Ariel has left.

   $19

37

Reteach
Order of Operations

To help you remember the order of operations use the phrase "Please Excuse My Dear Aunt Sally."

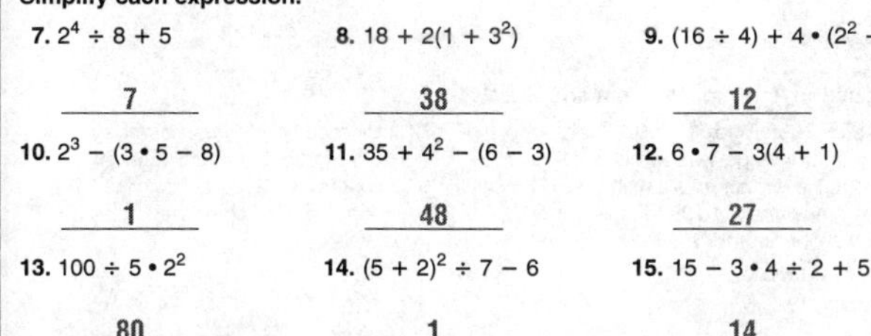

P: first, **p**arentheses (if any)
E: second, **e**xponents (if any)
M and **D:** then, **m**ultiplication and **d**ivision, in order from left to right
A and **S:** finally, **a**ddition and **s**ubtraction, in order from left to right

Evaluate. $39 \div (9 + 4) + 5 - 2^2$
   Parentheses → $39 \div 13 + 5 - 2^2$
   Exponents → $39 \div 13 + 5 - 4$
Multiply and divide from left to right → $3 + 5 - 4$
Add and subtract from left to right → $8 - 4 = 4$

Simplify each expression.

1. $12 \cdot 4 - 2$
   _48_ $- 2$
   _46_

2. $15 \div 3 \cdot 5$
   _5_ $\cdot 5$
   _25_

3. $15 \cdot 3 \div 5$
   _45_ $\div 5$
   _9_

4. $8 + 20 \div 4$
   _13_

5. $5 - 2 \cdot 6 \div 4 + 1$
   _3_

6. $3^2 + 6 \cdot 4 - 5^2$
   _8_

7. $1 + 4 \cdot 9 \div 6 - 7$
   _0_

8. $18 \div (6 \div 3)$
   _9_

9. $(18 \div 6) \div 3$
   _1_

10. $4 \cdot 5 + 8 \div 2 - 7$
   _17_

11. $2 \cdot 3 - 8 \div 2^2$
   _4_

12. $8(7 - 6) \div 2^3$
   _1_

38

Challenge
Fixed and Variable Costs

A *fixed cost* is a one-time cost. A *variable cost* changes depending on your use of a product or a service.

The annual enrollment fee per year at a fitness club (fixed cost) is $30. You also pay $2 per visit (variable cost), and you visit the club 8 times per month. What is your total annual cost?

$2 \cdot (8 \cdot 12)$	variable cost times total visits
$30 + 2 \cdot (8 \cdot 12)$	total annual cost
222	Your total annual cost is $222.

Use the information above to solve problems 1–4.

1. Suppose the annual fee is $25, but the cost per visit is $3. What is your annual cost?

   $313

2. Suppose you visit the club 3 times per week instead of 8 times per month. What is your annual cost?

   $342

3. Suppose the cost per visit is $3 after the first 50 visits per year. What is your annual cost?

   $268

4. Suppose you pay for up to 75 visits per year. Any additional visits are free. What is your annual cost?

   $180

The school band is raising money for a trip. The members ordered 5 dozen jerseys for $7 each and sold them for $12 each. They also ordered 4 dozen sweatshirts for $11 each and sold them for $18 each. The band paid $35 to create the design.

5. Write and simplify an expression to calculate the band's variable costs for the clothing.

   $7(5 \cdot 12) + 11(4 \cdot 12); \$948$

6. Write and simplify an expression to calculate the band's total cost, including fixed costs.

   $7(5 \cdot 12) + 11(4 \cdot 12) + 35; \$983$

7. Write and simplify an expression to calculate the band's profit.

   $12(5 \cdot 12) + 18(4 \cdot 12) - [7(5 \cdot 12) + 11(4 \cdot 12) + 35]; \$601$

8. What would the profit be if jerseys sold for $10 and sweatshirts for $20?

   $577

39
Holt Mathematics

Problem Solving
Order of Operations

Write the correct answer.

1. In 1975, the minimum wage was $2.10 per hour. Write and simplify an expression to show wages earned in a 35-hour week after a $12 tax deduction.

   $35 \cdot 2.10 - 12; \$61.50$

2. George bought 3 boxes of Girl Scout cookies at $3.50 per box and 4 boxes at $3.00 per box. Write and simplify an expression to show his total cost.

   $3 \cdot 3.50 + 4 \cdot 3; \$22.50$

3. In 1 week Ed works 4 days, 3 hours a day, for $12 per hour, and 2 days, 6 hours a day, for $15 per hour. Simplify the expression $12(4 \cdot 3) + 15(2 \cdot 6)$ to find Ed's weekly earnings.

   $324 per week

4. Keisha had $150. She bought jeans for $27, a sweater for $32, 3 blouses for $16 each, and 2 pairs of socks for $6 each. Simplify the expression $150 - [27 + 32 + (3 \cdot 16) + (2 \cdot 6)]$ to find out how much money she has left.

   She has $31 left.

Choose the letter for the best answer.

5. As of September 1, 1997, the minimum wage was set at $5.15 per hour. How much more would someone earn now than in 1997 if she earns $5 more per hour for a 40-hour week?

   A $206 more
   B $200 more
   C $406 more
   D $400 more

6. Gary received $200 in birthday gifts. He bought 5 CDs for $15 each, 2 posters for $12 each, and a $70 jacket. How much money does he have left?

   F $31
   G $10
   H $132
   J $169

7. Yvonne took her younger brother and his friends to the movies. She bought 5 tickets for $8 each, 4 drinks for $2 each, and two $3 containers of popcorn. How much did she spend?

   A $22
   B $51
   C $54
   D $38

8. On a business trip, Mr. Chang stayed in a hotel for 7 nights. He paid $149 per night. While he was there, he made 8 phone calls at $2 each and charged $81 to room service. How much did he spend?

   F $246
   G $946
   H $1,043
   J $1,140

40
Holt Mathematics

Reading Strategies
Use a Flowchart

When you read a book, you read from left to right. When you evaluate an expression, you cannot always work from left to right. You must follow a special rule called the **order of operations.** Use the flowchart below to help you follow the order of operations.

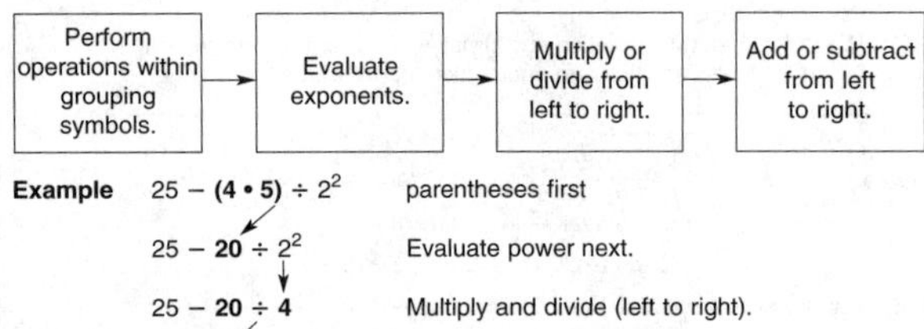

Example	$25 - (4 \cdot 5) \div 2^2$	parentheses first
	$25 - 20 \div 2^2$	Evaluate power next.
	$25 - 20 \div 4$	Multiply and divide (left to right).
	$25 - 5$	Add and subtract (left to right).
	20	

Answer each question.

1. In what order will you perform the operations in the following expression: $30 - 18 \div 2 \cdot 3 + 4$?

   divide ⟶ multiply ⟶ subtract ⟶ add

2. Simplify this expression: $30 - 18 \div 2 \cdot 3 + 4$. _____ 7 _____

3. Make a flow chart for the order of operations in this expression: $3^3 - (2 + 5 \cdot 2)$.

   parentheses ⟶ multiply ⟶ add ⟶ exponents ⟶ subtract

4. Simplify this expression: $3^3 - (2 + 5 \cdot 2)$. _____ 15 _____

5. Make a flow chart for the order of operations in this expression: $(3 + 12 \div 3)^2 - 4$.

   parentheses ⟶ divide ⟶ add ⟶ exponent ⟶ subtract

6. Simplify this expression: $(3 + 12 \div 3)^2 - 4$. _____ 45 _____

41
Holt Mathematics

Puzzles, Twisters & Teasers
Is Everything in Order?

What's the last thing you take off before you go to bed?

Decide whether each statement below is true or false. Use your answers to solve the riddle.

1. A numerical expression is made up of numbers and operations.
   (T)   F

2. In mathematics, as in life, tasks may be done in any order.
   T   (F)

3. When using the order of operations, you should do division after subtraction.
   T   (F)

4. When using the order of operations, you should subtract and add from left to right.
   (T)   F

5. When using the order of operations, you should divide and multiply from right to left.
   T   (F)

6. When an expression has a set of grouping symbols within a second set of grouping symbols, you should begin with the innermost set.
   (T)   F

7. You should perform operations inside parentheses first.
   (T)   F

8. When using the order of operations, you should evaluate the power expression after multiplying and adding.
   T   (F)

9. Mathematicians agree on using the order of operations.
   (T)   F

T
L F Y
R H

F
O T
E U

You take

Y	O	U	R	F	E	E	T	O	F	F
T	F	F	T	T	F	F	F	F	T	T

T	H	E	F	L	O	O	R
F	T	F	T	T	F	F	T

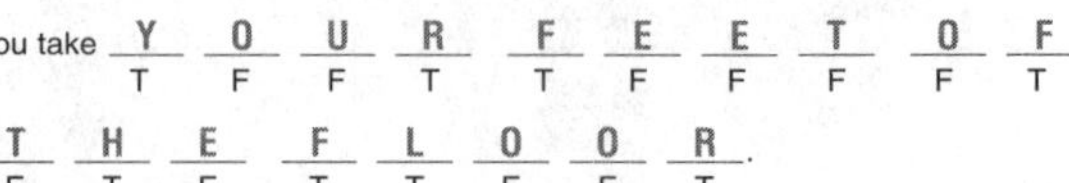

42
Holt Mathematics

108

Holt Mathematics

Tell which property is shown.

1. $5 + 0 = 1$

   Identity Property

2. $8 \cdot (6 \cdot 2) = (8 \cdot 6) \cdot 2$

   Associative Property

3. $9 + 8 = 8 + 9$

   Commutative Property

4. $4 \cdot 1 = 4$

   Identity Property

Simplify each expression. Write a reason for each step.

5. $13 + 28 + 7$

   $13 + 28 + 7 = 28 + 13 + 7$ Reason: Commutative Property

   $= 28 + (13 + 7)$ Reason: Associative Property

   $= 28 + \underline{20}$ Reason: Add.

   $= \underline{48}$ Reason: Add

6. $20 \cdot (17 \cdot 5)$

   $20 \cdot (17 \cdot 5) = 20 \cdot (\underline{5} \cdot 17)$ Reason: Commutative Property

   $= (20 \cdot \underline{5}) \cdot 17$ Reason: Associative Property

   $= \underline{100} \cdot \underline{17}$ Reason: Multiply.

   $= \underline{1,700}$ Reason: Multiply

Use the Distributive Property to find each product.

7. $4(17)$

   $4(17) = 4 \cdot (10 + \underline{7})$

   $= (4 \cdot \underline{10}) + (4 \cdot 7)$

   $= \underline{40} + \underline{28}$

   $= \underline{68}$

8. $3(28)$ Possible work shown.

   $3(28) = \underline{3 \cdot (20 + 8)}$

   $= \underline{(3 \cdot 20) + (3 \cdot 8)}$

   $= \underline{60 + 24}$

   $= \underline{84}$

    43   **Holt Mathematics**

Tell which property is represented.

1. $12 \cdot 14 = 14 \cdot 12$

   Commutative Property

2. $1 \cdot 36 = 36$

   Identity Property

3. $(17 + 36) + 4 = 17 + (36 + 4)$

   Associative Property

4. $8 \cdot 12 \cdot 5 = 8 \cdot (12 \cdot 5)$

   Associative Property

Simplify each expression. Justify each step.

5. $4 \cdot 9 \cdot 50$ Possible answers are given.

   $4 \cdot 9 \cdot 50 = \underline{9 \cdot 4 \cdot 50};$ Commutative Property

   $= \underline{9 \cdot (4 \cdot 50)};$ Associative Property

   $= \underline{9 \cdot 200};$ Multiply.

   $= \underline{1,800};$ Multiply.

6. $(33 + 45) + 7$ Possible answers are given.

   $(33 + 45) + 7 = \underline{(45 + 33) + 7};$ Commutative Property

   $= \underline{45 + (33 + 7)};$ Associative Property

   $= \underline{45 + 40};$ Add.

   $= \underline{85};$ Add.

Use the Distributive Property to find each product. Possible work shown.

7. $3(26) = \underline{3 \cdot (20 + 6)}$

   $= \underline{(3 \cdot 20) + (3 \cdot 6)}$

   $= \underline{60 + 18}$

   $= \underline{78}$

8. $(18)9 = \underline{(20 - 2)9}$

   $= \underline{(20 \cdot 9) - (2 \cdot 9)}$

   $= \underline{180 - 18}$

   $= \underline{162}$

    44   **Holt Mathematics**

Complete each equation. Then tell which property is represented.

1. $23 + \underline{0} = 23$

   Identity Property

2. $\underline{7} \cdot (19 + 6) = (7 \cdot 19) + (7 \cdot 6)$

   Distributive Property

3. $27 + 45 = \underline{45} + 27$

   Commutative Property

4. $6 \cdot (\underline{14} \cdot 7) = (6 \cdot 14) \cdot 7$

   Associative Property

Simplify each expression. Justify each step.

5. $(40 \cdot 7) \cdot 5$ Possible answers are given.

   $(40 \cdot 7) \cdot 5 = \underline{(7 \cdot 40) \cdot 5};$ Commutative Property

   $= \underline{7 \cdot (40 \cdot 5)};$ Associative Property

   $= \underline{7 \cdot 200};$ Multiply.

   $= \underline{1,400};$ Multiply.

6. $15 + 98 + 85$ Possible answers are given.

   $15 + 98 + 85 = \underline{98 + 15 + 85};$ Commutative Property

   $= \underline{98 + (15 + 85)};$ Associative Property

   $= \underline{98 + 100};$ Add.

   $= \underline{198}$

Use the Distributive Property to find each product. Possible work shown.

7. $7(43) = \underline{7 \cdot (40 + 3)}$

   $= \underline{(7 \cdot 40) + (7 \cdot 3)}$

   $= \underline{280 + 21}$

   $= \underline{301}$

8. $(597)4 = \underline{(600 - 3)4}$

   $= \underline{(600 \cdot 4) - (3 \cdot 4)}$

   $= \underline{2,400 - 12}$

   $= \underline{2,388}$

    45   **Holt Mathematics**

You can use the Commutative Property, the Associative Property, and
the Distributive Property with mental math to simplify expressions.

$16 + 47 + 14 = \mathbf{47 + 16 + 14}$ Commutative Property	$8 \cdot 3 \cdot 5 = 3 \cdot 8 \cdot 5$
$= 47 + (16 + 14)$ Associative Property	$= 3 \cdot (8 \cdot 5)$
$= 47 + 40$ Mental math	$= 3 \cdot 40$
$= 87$ Mental math	$= 120$
$9(28) = 9(20 + 8)$	$9(28) = 9(30 - 2)$
$= (9 \cdot 20) + (9 \cdot 8)$ Distributive Property	$= (9 \cdot 30) - (9 \cdot 2)$
$= 180 + 72$ Mental math	$= 270 - 18$
$= 252$ Mental math	$= 252$

Simplify each expression. Tell what properties you used.

1. $(45 + 39) + 25 = (39 + \underline{45}) + 25$ **Commutative** Property

   $= 39 + (\underline{45} + \underline{25})$ **Associative** Property

   $= 39 + \underline{70}$

   $= \underline{109}$

2. $25 \cdot 7 \cdot 4 = 25 \cdot \underline{4} \cdot \underline{7}$ **Commutative** Property

   $= (\underline{25} \cdot \underline{4}) \cdot 7$ **Associative** Property

   $= \underline{100} \cdot \underline{7}$

   $= \underline{700}$

3. $3.5(18) = 5 \cdot (10 + \underline{8})$

   $= (5 \cdot \underline{10}) + (5 \cdot \underline{8})$

   $= \underline{50} + \underline{40}$

   $= \underline{90}$

   Distributive Property

4. $6(29) = 6 \cdot (30 - \underline{1})$

   $= (6 \cdot \underline{30}) - (6 \cdot \underline{1})$

   $= \underline{180} - \underline{6}$

   $= \underline{174}$

   Distributive Property

    46   **Holt Mathematics**

Challenge
What's the Expression?

On each tic-tac-toe board below, exactly one row, column, or diagonal contains three expressions with the same value.

Predict which row, column, or diagonal of squares will have three expressions with the same value. Draw a line through those three squares. Then check your prediction. Find the value of each expression on the board.

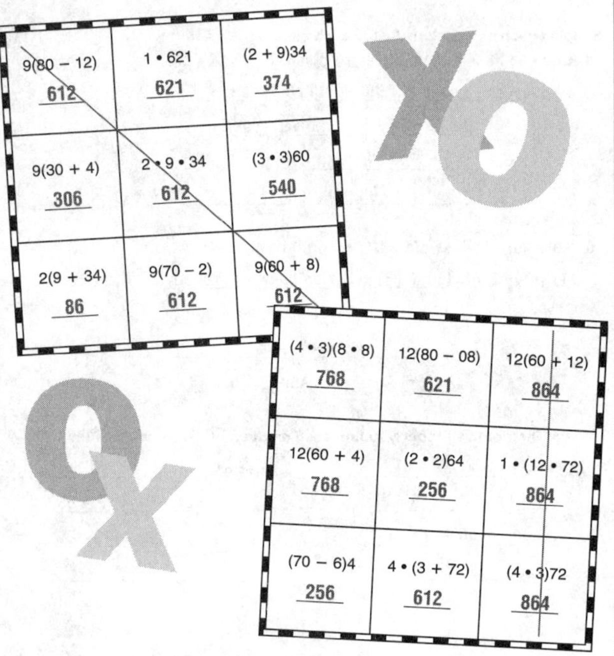

47   **Holt Mathematics**

Problem Solving
Properties

Write the correct answer.

1. Jo makes and sells jewelry. She sold three bracelets for $45, $17, and $25. Write an expression for the total Jo received. Explain how you can use properties and mental math to simplify the expression.

Possible answer: $45 + $17 + $25; use the commutative and associative properties to rewrite the expression as ($45 + $25) + $17, then add mentally.

2. Use parentheses to show two ways of grouping the numbers in 12 • 8 • 25. Tell which expression you think would be easier to simplify, and why. Then simplify the expression.

Possible answer: (12 • 8) • 25, 12 • (8 • 25); 12 • (8 • 25) is easier to multiply, because 12 • 200 is easier to multiply than 96 • 25; 2,400

3. The distance from Mark's apartment to his job is 27 miles. Mark works 5 days per week. How many miles does Mark drive to and from work each week?

270 mi

4. Jane said that 6(64) = 6(50) + 6(14). Is she correct? Use the Distributive Property to explain your answer.

Possible answer: Yes. Since 50 + 14 = 64, 6(64) = 6(50 + 14) = 6(50) + 6 • 14).

Choose the letter for the best answer.

5. Maxine works 8 hours at a rate of $16 per hour. Which expression could **not** be used to find her total earnings in dollars?

(A) 8 • (10 • 6)
B 8 • (20 − 4)
C 8 • (10 + 6)
D 8 • (8 + 8)

6. Rosemary runs 16 miles on Friday, 8 miles on Saturday, and 14 miles on Sunday. How many miles does Rosemary run in all?

F 22 mi
G 24 mi
H 30 mi
(J) 38 mi

7. Which of the following represents the Identity Property?

A (8 • 4) • 3 = 8 • (4 • 3)
B 16 • 0 = 0
(C) 25 • 1 = 25
D 6(26) = 6(20) + 6(6)

8. Which of the following shows how the Distributive Property could be used to simplify 7(28)?

F 7 • 2 • 8
G 7 • (20 • 8)
(H) 7 • (20 + 8)
J (7 • 20) + 8

48   **Holt Mathematics**

Reading Strategies
Use a Flowchart

Use a flowchart to help you simplify an expression, such as (25 + 89) + 15.

Step 1: Choose two numbers that are easy to add.
(25 + 89) + 15

Step 2: Rewrite the expression so the two numbers are next to each other.
Use the Commutative Property. (25 + 89) + 15 = (89 + 25) + 15)

Step 3: Rewrite the expression so the two numbers are grouped together.
Use the Associative Property. (89 + 25) + 15 = 89 + (25 + 15)

Step 4: Add.
89 + (25 + 15) = 89 + 40 = 129

Use the expression 16 + (39 + 14) for Exercises 1–4.

1. Which two numbers are easy to add?   16 and 14

2. Rewrite the expression so that the numbers that are easy to add are next to each other. What property lets you do this?

16 + (14 + 39); Commutative Property

3. Rewrite the expression so that the numbers that are easy to add are grouped together. What property lets you do this?

(16 + 14) + 39; Associative Property

4. Simplify the expression.
69

Use the expression 35 + 47 + 5 for Exercises 5–8.

5. Which two numbers are easy to add?
35 and 5

6. Rewrite the expression so that the numbers that are easy to add are next to each other. What property lets you do this?

47 + 35 + 5; Commutative Property

7. Rewrite the expression so that the numbers that are easy to add are grouped together. What property lets you do this?

47 + (35 + 5); Associative Property

8. Simplify the expression.
87

49   **Holt Mathematics**

Puzzles, Twisters, & Teasers
Make the Connection!

Draw a line to connect each equation to the property it represents.

1. 3 • (8 • 5) = (3 • 8) • 5
2. 1 • 15 = 15
3. 9(4 + 6) = 9(4) + 9(6)
4. 38 + 7 = 7 + 38

Identity Property   R
Commutative Property   Y
Associative Property   E
Distributive Property   P

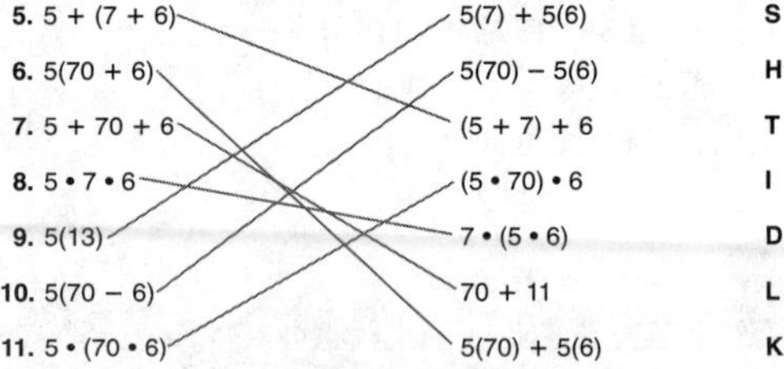

Draw line to connect expressions with the same value.

5. 5 + (7 + 6)
6. 5(70 + 6)
7. 5 + 70 + 6
8. 5 • 7 • 6
9. 5(13)
10. 5(70 − 6)
11. 5 • (70 • 6)

5(7) + 5(6)   S
5(70) − 5(6)   H
(5 + 7) + 6   T
(5 • 70) • 6   I
7 • (5 • 6)   D
70 + 11   L
5(70) + 5(6)   K

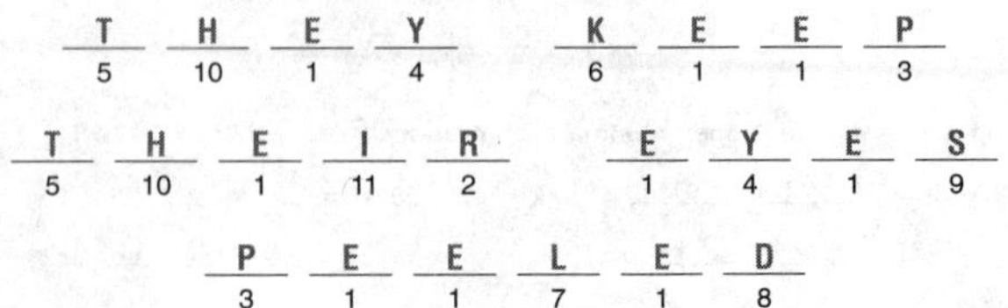

Start with number 1. Find the letter next to the answer. Use the letters to find out why potatoes are good detectives.

T	H	E	Y		K	E	E	P
5	10	1	4		6	1	1	3

T	H	E	I	R		E	Y	E	S
5	10	1	11	2		1	4	1	9

P	E	E	L	E	D
3	1	1	7	1	8

50   **Holt Mathematics**

    Holt Mathematics

Practice A
Variables and Algebraic Expressions

Find the value of $n + 3$ for each value of n.

1. $n = 4$ **2.** $n = 7$ **3.** $n = 0$ **4.** $n = 32$

   7   10   3   35

Find the value of $x - 9$ for each value of x.

5. $x = 12$ **6.** $x = 57$ **7.** $x = 19$ **8.** $x = 100$

   3   48   10   91

Find the value of each expression using the given value for each variable.

9. $3n$ for $n = 4$ **10.** $x + 8$ for $x = 8$ **11.** $9p - 6$ for $p = 2$

   12   16   12

12. $n \div 5$ for $n = 35$ **13.** $6x + 18$ for $x = 0$ **14.** $s - 7$ for $s = 8$

   7   18   1

15. $3w + 5$ for $w = 3$ **16.** $c - 9$ for $c = 12$ **17.** $2a \div 3$ for $a = 6$

   14   3   4

18. $y + z$ for $y = 10$ and $z = 20$ **19.** $3w - 2v$ for $w = 7$ and $v = 8$

   30   5

20. $4a \div b$ for $a = 6$ and $b = 4$ **21.** $5s + 4t$ for $s = 3$ and $t = 4$

   6   31

22. The expression $7w$ gives the number of days in w weeks. Find the value of $7w$ for $w = 20$. How many days are there in 20 weeks?

   140 days

23. A cat can run as fast as $m \div 2$ miles per minute in m minutes. Find the value of $m \div 2$ for $m = 10$. How many miles can a cat run in 10 minutes?

   5 miles

24. Tyrone works 8 hours a day. You can use the expression $8d$ to find the total number of hours he works in d days. How many hours does he work in 5 days?

   40 hours

Holt Mathematics

Practice B
Variables and Algebraic Expressions

Evaluate $n - 5$ for each value of n.

1. $n = 8$ **2.** $n = 121$ **3.** $n = 32$ **4.** $n = 59$

   3   116   27   54

Evaluate each expression for the given values of the variable.

5. $3n + 15$ for $n = 4$ **6.** $h \div 12$ for $h = 60$ **7.** $32x - 32$ for $x = 2$

   27   5   32

8. $\frac{c}{2}$ for $c = 24$ **9.** $(n \div 2)5$ for $n = 14$ **10.** $8p + 148$ for $p = 15$

   12   35   268

11. $e^2 - 7$ for $e = 8$ **12.** $3d^2 + d$ for $d = 5$ **13.** $40 - 4k^3$ for $k = 2$

   57   80   8

14. $2y - z$ for $y = 21$ and $z = 19$ **15.** $3h^2 + 8m$ for $h = 3$ and $m = 2$

   23   43

16. $18 \div a + b \div 9$ for $a = 6$ and $b = 45$ **17.** $10x - 4y$ for $x = 14$ and $y = 5$

   8   120

18. You can find the area of a rectangle with the expression lw where l represents the length and w represents the width. What is the area of the rectangle at right in square feet?   5 ft   2 ft

   10 square feet

19. Rita drove an average of 55 mi/h on her trip to the mountains. You can use the expression $55h$ to find out how many miles she drove in h hours. If she drove for 5 hours, how many miles did she drive?

   275 miles

Holt Mathematics

Practice C
Variables and Algebraic Expressions

Evaluate each expression for the given values of the variable.

1. $3n + 4n$ for $n = 8$ **2.** $\frac{6s}{5}$ for $s = 25$

   56   30

3. $q^2 + 5q - 11$ for $q = 4$ **4.** $\frac{350}{d} + 4d + 7$ for $d = 10$

   25   82

5. $9x + 2x^2 + 2$ for $x = 2$ **6.** $8m^2 + 7 - 2m$ for $m = 3$

   28   73

7. $4(h + k)$ for $h = 3$ and $k = 55$ **8.** $\frac{6r}{4} + 5s$ for $r = 8$ and $s = 18$

   232   102

9. $6a - 2b^2$ for $a = 9$ and $b = 5$ **10.** $6h - 20g$ for $h = 1{,}500$ and $g = 200$

   4   5,000

11. $\frac{36}{m^2} + \frac{n^2}{4}$ for $m = 6$ and $n = 16$ **12.** $x^2 - 2x - y^2$ for $x = 15$ and $y = 2$

   65   191

13. $4d^3 + 6e^2 - \frac{8d}{2}$ for $d = 2$ and $e = 3$ **14.** $\frac{5r^2}{4} + \frac{4s}{3r}$ for $r = 4$ and $s = 9$

   78   23

15. You can find the volume of a rectangular prism with the expression $a \bullet b \bullet c$, where a is the length, b is the width, and c is the height of the prism. What is the volume of the prism at the right in cubic inches?   8 in.   3 in.   2 in.

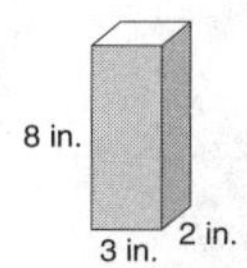

   48 cubic inches

16. You can use the expression $5m$ to find out how many seconds it takes a sound to travel m miles through the air. However, through water, sound takes m seconds to travel m miles. Use the expression $5m - m$ to find out how much longer will it take a sound to travel 8 miles in air than in water.

   32 seconds longer

Holt Mathematics

Reteach
Variables and Algebraic Expressions

A **variable** is a letter that represents a number than can change in an expression. When you **evaluate** an algebraic expression, you substitute the value given for the variable in the expression.

- Algebraic expression: $x - 3$

  The value of the expression depends on the value of the variable x.

  If $x = 7 \rightarrow 7 - 3 = 4$

  If $x = 11 \rightarrow 11 - 3 = 8$

  If $x = 15 \rightarrow 15 - 3 = 12$

- Evaluate $4n + 1$ for $n = 5$.

  Replace the variable n with 5. $\rightarrow 4(5) + 1 = 20 + 1 = 21$

Evaluate each expression for the given value.

1. $a + 7$ for $a = 3$ **2.** $k - 5$ for $k = 13$

   $a + 7 = 3 + 7 = $ __10__ $k - 5 = $ __13__ $- 5 = $ __8__

3. $y \div 3$ for $y = 6$ **4.** $12 + m$ for $m = 9$

   $y \div 3 = $ __6__ $\div 3 = $ __2__ $12 + m = $ __12__ $+ $ __9__ $= $ __21__

5. $3n - 2$ for $n = 5$

   $3n - 2 = 3($ __5__ $) - 2 = $ __15__ $- 2 = $ __13__

6. $5x + 4$ for $x = 4$

   $5x + 4 = 5($ __4__ $) + $ __4__ $= $ __20__ $+ $ __4__ $= $ __24__

7. $c - 9$ for $c = 11$ **8.** $b + 16$ for $b = 4$ **9.** $a - 4$ for $a = 9$

   2   20   5

10. $25 - g$ for $g = 12$ **11.** $w + 5$ for $w = 2$ **12.** $3 + s$ for $s = 8$

   13   7   11

13. $7q$ for $q = 10$ **14.** $2y + 9$ for $y = 8$ **15.** $6x - 3$ for $x = 1$

   70   25   3

Holt Mathematics

Holt Mathematics

Challenge
What an Expression!

Complete each table with four expressions that have the same value. Use each given value of *n*.

1.

Expression Value 32	Value of *n* 4
Addition expression: $n + 28$	
Subtraction expression: $36 - n$	
Multiplication expression: $8n$	
Division expression: $128 \div n$	

2.

Expression Value 96	Value of *n* 12
Addition expression: $n + 84$	
Subtraction expression: $108 - n$	
Multiplication expression: $8n$	
Division expression: $1{,}152 \div n$	

3.

Expression Value 156	Value of *n* 6
Addition expression: $n + 150$	
Subtraction expression: $162 - n$	
Multiplication expression: $26n$	
Division expression: $936 \div n$	

4.

Expression Value 98	Value of *n* 14
Addition expression: $n + 84$	
Subtraction expression: $112 - n$	
Multiplication expression: $7n$	
Division expression: $1{,}372 \div n$	

5.

Expression Value 57	Value of *n* 12
Addition expression: $n + 45$	
Subtraction expression: $69 - n$	
Multiplication expression: $4.75n$	
Division expression: $684 \div n$	

6.

Expression Value 248	Value of *n* 124
Addition expression: $n + 124$	
Subtraction expression: $372 - n$	
Multiplication expression: $2n$	
Division expression: $30{,}752 \div n$	

55   **Holt Mathematics**

Problem Solving
Variables and Expressions

Write the correct answer.

1. In 2000, people in the United States watched television an average of 29 hours per week. Use the expression $29w$ for $w = 4$ to find out about how many hours per month this is.

about 116 hours

2. Find the value of the variable *w* in the expression $29w$ to find the average number of hours people watched television in a year. Find the value of the expression.

52; 1,508 hours

3. The expression $y + 45$ gives the year when a person will be 45 years old, where *y* is the year of birth. When will a person born in 1992 be 45 years old?

2037

4. The expression $24g$ gives the number of miles Guy's car can travel on *g* gallons of gas. If the car has 6 gallons of gas left, how much farther can he drive?

144 miles

Choose the letter for the best answer.

5. Sam is 5 feet tall. The expression $0.5m + 60$ can be used to calculate his height in inches if he grows an average of 0.5 inch each month. How tall will Sam be in 6 months?

A 56 inches
B 5 feet 6 inches
C 63 inches
D 53 inches

6. The winner of the 1911 Indianapolis 500 auto race drove at a speed of about $s - 56$ mi/h, where *s* is the 2001 winning speed of about 131 mi/h. What was the approximate winning speed in 1911?

F 75 mi/h
G 186 mi/h
H 85 mi/h
J 187 mi/h

7. The expression $1{,}587v$ gives the number of pounds of waste produced per person in the United States in *v* years. How many pounds of waste per person is produced in the United States in 6 years?

A 1,581 pounds
B 1,593 pounds
C 9,348 pounds
D 9,522 pounds

8. The expression $\$1.25p + \$3.50$ can be used to calculate the total charge for faxing *p* pages at a business services store. How much would it cost to fax 8 pages?

F $12.50
G $4.75
H $13.50
J $10.00

56   **Holt Mathematics**

Reading Strategies
Focus on Vocabulary

Your age varies from one year to the next. The cost of gasoline can vary from one week to the next. The word **vary** means change. In mathematical expressions, the value of a letter can change, or vary, so letters are called **variables.**

The opposite of vary is to stay **constant.** A constant value never changes.

An algebraic expression is made up of constants, variables, and operation symbols.

> algebraic expression = constants + variables + operation symbols
>
> Examples: $4x + 3y - 2^2$ $6p - 3t + 12$

To evaluate an algebraic expression, you need to:

Step 1: Substitute a value for each variable. → **Step 2:** Follow the order of operations.

Evaluate: $4a - 2b + 8$ for $a = 5$ and $b = 3$.

$4(5) - 2(3) + 8$	Substitute 5 for *a* and 3 for *b*.
$20 - 6 + 8$	Multiply first.
$14 + 8$	Add and subtract from left to right.
22	

Use this expression for Exercises 1–7: $8 + 4p - 2t$.

1. Write the variables in this expression. _______ *p* and *t*

2. Rewrite the expression by substituting these values: $p = 6$ and $t = 2$.

$8 + 4(6) - 2(2)$

3. What operation will you perform first? _______ multiplication

4. Perform this operation and rewrite the expression.

$8 + 24 - 4$

5. What operation will you perform next? _______ addition

6. Perform this operation and rewrite the expression. _______ $32 - 4$

57   **Holt Mathematics**

Puzzles, Twisters & Teasers
Movie Math!

Circle words from the list in the word search. Then find an extra word in the word search that best completes the riddle.

expression   substitute   variable   value
constant   evaluate   algebra   algebraic

This Ron Howard movie was all wet.

S   P   L   A   S   H

58   **Holt Mathematics**

112

Holt Mathematics

LESSON 1-7 — Practice A
Variables and Algebraic Expressions

Find the value of $n + 3$ for each value of n.

1. $n = 4$ 2. $n = 7$ 3. $n = 0$ 4. $n = 32$
   7   10   3   35

Find the value of $x - 9$ for each value of x.

5. $x = 12$ 6. $x = 57$ 7. $x = 19$ 8. $x = 100$
   3   48   10   91

Find the value of each expression using the given value for each variable.

9. $3n$ for $n = 4$ 10. $x + 8$ for $x = 8$ 11. $9p - 6$ for $p = 2$
   12   16   12

12. $n \div 5$ for $n = 35$ 13. $6x + 18$ for $x = 0$ 14. $s - 7$ for $s = 8$
   7   18   1

15. $3w + 5$ for $w = 3$ 16. $c - 9$ for $c = 12$ 17. $2a \div 3$ for $a = 6$
   14   3   4

18. $y + z$ for $y = 10$ and $z = 20$ 19. $3w - 2v$ for $w = 7$ and $v = 8$
   30   5

20. $4a \div b$ for $a = 6$ and $b = 4$ 21. $5s + 4t$ for $s = 3$ and $t = 4$
   6   31

22. The expression $7w$ gives the number of days in w weeks. Find the value of $7w$ for $w = 20$. How many days are there in 20 weeks?   **140 days**

23. A cat can run as fast as $m \div 2$ miles per minute in m minutes. Find the value of $m \div 2$ for $m = 10$. How many miles can a cat run in 10 minutes?   **5 miles**

24. Tyrone works 8 hours a day. You can use the expression $8d$ to find the total number of hours he works in d days. How many hours does he work in 5 days?   **40 hours**

    51   **Holt Mathematics**

LESSON 1-7 — Practice B
Variables and Algebraic Expressions

Evaluate $n - 5$ for each value of n.

1. $n = 8$ 2. $n = 121$ 3. $n = 32$ 4. $n = 59$
   3   116   27   54

Evaluate each expression for the given values of the variable.

5. $3n + 15$ for $n = 4$ 6. $h \div 12$ for $h = 60$ 7. $32x - 32$ for $x = 2$
   27   5   32

8. $\frac{c}{2}$ for $c = 24$ 9. $(n \div 2)5$ for $n = 14$ 10. $8p + 148$ for $p = 15$
   12   35   268

11. $e^2 - 7$ for $e = 8$ 12. $3d^2 + d$ for $d = 5$ 13. $40 - 4k^3$ for $k = 2$
   57   80   8

14. $2y - z$ for $y = 21$ and $z = 19$ 15. $3h^2 + 8m$ for $h = 3$ and $m = 2$
   23   43

16. $18 \div a + b \div 9$ for $a = 6$ and $b = 45$ 17. $10x - 4y$ for $x = 14$ and $y = 5$
   8   120

18. You can find the area of a rectangle with the expression lw where l represents the length and w represents the width. What is the area of the rectangle at right in square feet?   5 ft   2 ft

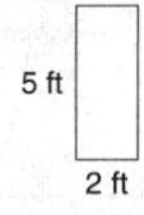

   10 square feet

19. Rita drove an average of 55 mi/h on her trip to the mountains. You can use the expression $55h$ to find out how many miles she drove in h hours. If she drove for 5 hours, how many miles did she drive?

   275 miles

    52   **Holt Mathematics**

LESSON 1-7 — Practice C
Variables and Algebraic Expressions

Evaluate each expression for the given values of the variable.

1. $3n + 4n$ for $n = 8$ 2. $\frac{6s}{5}$ for $s = 25$
   56   30

3. $q^2 + 5q - 11$ for $q = 4$ 4. $\frac{350}{d} + 4d + 7$ for $d = 10$
   25   82

5. $9x + 2x^2 + 2$ for $x = 2$ 6. $8m^2 + 7 - 2m$ for $m = 3$
   28   73

7. $4(h + k)$ for $h = 3$ and $k = 55$ 8. $\frac{6r}{4} + 5s$ for $r = 8$ and $s = 18$
   232   102

9. $6a - 2b^2$ for $a = 9$ and $b = 5$ 10. $6h - 20g$ for $h = 1{,}500$ and $g = 200$
   4   5,000

11. $\frac{36}{m^2} + \frac{n^2}{4}$ for $m = 6$ and $n = 16$ 12. $x^2 - 2x - y^2$ for $x = 15$ and $y = 2$
   65   191

13. $4d^3 + 6e^2 - \frac{8d}{2}$ for $d = 2$ and $e = 3$ 14. $\frac{5r^2}{4} + \frac{4s}{3r}$ for $r = 4$ and $s = 9$
   78   23

15. You can find the volume of a rectangular prism with the expression $a \cdot b \cdot c$, where a is the length, b is the width, and c is the height of the prism. What is the volume of the prism at the right in cubic inches?   8 in.   3 in.   2 in.

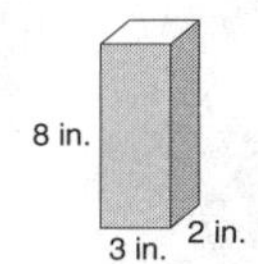

   48 cubic inches

16. You can use the expression $5m$ to find out how many seconds it takes a sound to travel m miles through the air. However, through water, sound takes m seconds to travel m miles. Use the expression $5m - m$ to find out how much longer will it take a sound to travel 8 miles in air than in water.

   32 seconds longer

    53   **Holt Mathematics**

LESSON 1-7 — Reteach
Variables and Algebraic Expressions

A **variable** is a letter that represents a number than can change in an expression. When you **evaluate** an algebraic expression, you substitute the value given for the variable in the expression.

- Algebraic expression: $x - 3$

  The value of the expression depends on the value of the variable x.

  If $x = 7 \rightarrow 7 - 3 = 4$

  If $x = 11 \rightarrow 11 - 3 = 8$

  If $x = 15 \rightarrow 15 - 3 = 12$

- Evaluate $4n + 1$ for $n = 5$.

  Replace the variable n with 5. $\rightarrow 4(5) + 1 = 20 + 1 = 21$

Evaluate each expression for the given value.

1. $a + 7$ for $a = 3$ 2. $k - 5$ for $k = 13$
   $a + 7 = 3 + 7 =$ **10** $k - 5 =$ **13** $- 5 =$ **8**

3. $y \div 3$ for $y = 6$ 4. $12 + m$ for $m = 9$
   $y \div 3 =$ **6** $\div 3 =$ **2** $12 + m =$ **12** $+$ **9** $=$ **21**

5. $3n - 2$ for $n = 5$
   $3n - 2 = 3($ **5** $) - 2 =$ **15** $- 2 =$ **13**

6. $5x + 4$ for $x = 4$
   $5x + 4 = 5($ **4** $) +$ **4** $=$ **20** $+$ **4** $=$ **24**

7. $c - 9$ for $c = 11$ 8. $b + 16$ for $b = 4$ 9. $a - 4$ for $a = 9$
   2   20   5

10. $25 - g$ for $g = 12$ 11. $w + 5$ for $w = 2$ 12. $3 + s$ for $s = 8$
   13   7   11

13. $7q$ for $q = 10$ 14. $2y + 9$ for $y = 8$ 15. $6x - 3$ for $x = 1$
   70   25   3

    54   **Holt Mathematics**

Challenge

1-7 What an Expression!

Complete each table with four expressions that have the same
value. Use each given value of n.

1.

Expression Value 32	Value of n 4
Addition expression: $n + 28$	
Subtraction expression: $36 - n$	
Multiplication expression: $8n$	
Division expression: $128 \div n$	

2.

Expression Value 96	Value of n 12
Addition expression: $n + 84$	
Subtraction expression: $108 - n$	
Multiplication expression: $8n$	
Division expression: $1,152 \div n$	

3.

Expression Value 156	Value of n 6
Addition expression: $n + 150$	
Subtraction expression: $162 - n$	
Multiplication expression: $26n$	
Division expression: $936 \div n$	

4.

Expression Value 98	Value of n 14
Addition expression: $n + 84$	
Subtraction expression: $112 - n$	
Multiplication expression: $7n$	
Division expression: $1,372 \div n$	

5.

Expression Value 57	Value of n 12
Addition expression: $n + 45$	
Subtraction expression: $69 - n$	
Multiplication expression: $4.75n$	
Division expression: $684 \div n$	

6.

Expression Value 248	Value of n 124
Addition expression: $n + 124$	
Subtraction expression: $372 - n$	
Multiplication expression: $2n$	
Division expression: $30,752 \div n$	

55   Holt Mathematics

Problem Solving

1-7 Variables and Expressions

Write the correct answer.

1. In 2000, people in the United States watched television an average of 29 hours per week. Use the expression $29w$ for $w = 4$ to find out about how many hours per month this is.

about 116 hours

2. Find the value of the variable w in the expression $29w$ to find the average number of hours people watched television in a year. Find the value of the expression.

52; 1,508 hours

3. The expression $y + 45$ gives the year when a person will be 45 years old, where y is the year of birth. When will a person born in 1992 be 45 years old?

2037

4. The expression $24g$ gives the number of miles Guy's car can travel on g gallons of gas. If the car has 6 gallons of gas left, how much farther can he drive?

144 miles

Choose the letter for the best answer.

5. Sam is 5 feet tall. The expression $0.5m + 60$ can be used to calculate his height in inches if he grows an average of 0.5 inch each month. How tall will Sam be in 6 months?

A 56 inches
B 5 feet 6 inches
C 63 inches
D 53 inches

6. The winner of the 1911 Indianapolis 500 auto race drove at a speed of about $s - 56$ mi/h, where s is the 2001 winning speed of about 131 mi/h. What was the approximate winning speed in 1911?

F 75 mi/h
G 186 mi/h
H 85 mi/h
J 187 mi/h

7. The expression $1,587v$ gives the number of pounds of waste produced per person in the United States in v years. How many pounds of waste per person is produced in the United States in 6 years?

A 1,581 pounds
B 1,593 pounds
C 9,348 pounds
D 9,522 pounds

8. The expression $\$1.25p + \$3.50$ can be used to calculate the total charge for faxing p pages at a business services store. How much would it cost to fax 8 pages?

F $12.50
G $4.75
H $13.50
J $10.00

56   Holt Mathematics

Reading Strategies

1-7 Focus on Vocabulary

Your age varies from one year to the next. The cost of gasoline can vary from one week to the next. The word **vary** means change. In mathematical expressions, the value of a letter can change, or vary, so letters are called **variables.**

The opposite of vary is to stay **constant.** A constant value never changes.

An algebraic expression is made up of constants, variables, and operation symbols.

> algebraic expression = constants + variables + operation symbols
>
> Examples: $4x + 3y - 2^2$ $6p - 3t + 12$

To evaluate an algebraic expression, you need to:

Step 1: Substitute a value for each variable. → **Step 2:** Follow the order of operations.

Evaluate: $4a - 2b + 8$ for $a = 5$ and $b = 3$.

$4(5) - 2(3) + 8$ —— Substitute 5 for a and 3 for b.
$20 - 6 + 8$ —— Multiply first.
$14 + 8$ —— Add and subtract from left to right.
22

Use this expression for Exercises 1–7: $8 + 4p - 2t$.

1. Write the variables in this expression. _______ p and t

2. Rewrite the expression by substituting these values: $p = 6$ and $t = 2$.

$8 + 4(6) - 2(2)$

3. What operation will you perform first? _______ multiplication

4. Perform this operation and rewrite the expression.

$8 + 24 - 4$

5. What operation will you perform next? _______ addition

6. Perform this operation and rewrite the expression. _______ $32 - 4$

57   Holt Mathematics

Puzzles, Twisters & Teasers

1-7 Movie Math!

Circle words from the list in the word search. Then find an extra word in the word search that best completes the riddle.

expression   substitute   variable   value
constant   evaluate   algebra   algebraic

```
V A L U E S D E X P R E S S I O N
Q A C V B N M V A S D F U B H U I
W L R L K J G A C T U P B V M A L
E G G I S P L A S H W S I K L V
R E C V A R T U M J U I T C V G N
T B N H T B O A S D F G I P O E U
Y R X O P N L T E R W T T Y T B E
U A V G T M K E U H B J U W Q R D
O I P O I U Y T F I J O T A X A G
P C O N S T A N T D F G E Z X C V
```

This Ron Howard movie was all wet.

S   P   L   A   S   H

58   Holt Mathematics

Write as an algebraic expression.

1. the sum of m and 8

   $m + 8$

2. the product of 3 and n

   $3n$

3. 4 less than x

   $x - 4$

4. the quotient of a number and 12

   $a \div 12$

5. 52 times a number

   $52k$

6. w less than 15

   $15 - w$

7. the sum of 13 and a number

   $13 + e$

8. the product of 5 and p, increased by 10

   $5p + 10$

9. the sum of 15 divided by b and 6

   $15 \div b + 6$

10. 12 less than the amount y divided by 2

    $y \div 2 - 12$

11. 26 increased by 12 times a number $26 + 12s$

12. the difference of 2 times a number and 6 $2a - 6$

13. the product of h and 3, increased by 20 $3h + 20$

14. 18 less than the product of a number and 4 $4n - 18$

15. take away 32 from the product of 6 and a number $6z - 32$

16. Used video games cost \$25 each. Write an algebraic expression to find the cost of m video games. $25m$

17. Sal earned \$740 for n weeks of work. Write an algebraic expression for the amount he earned each week. $740 \div n$

18. At the end of the 2004–2005 NBA season, Reggie Miller was the all-time leader in 3-point field goals made. He made n more field goals than Dale Ellis. Dale Ellis made 1,719 3-pointers. Write an algebraic expression to find the number of 3-pointers Reggie Miller made. $1{,}719 + n$

19. The \$2 bill has Thomas Jefferson on the front of it. Write an algebraic expression to find out how much money v bills with Thomas Jefferson on them would be worth. $2v$

59
Holt Mathematics

Write each phrase as an algebraic expression.

1. 125 decreased by a number

   $125 - n$

2. 359 more than z

   $z + 359$

3. the product of a number and 35

   $35f$

4. the quotient of 100 and w

   $100 \div w$

5. twice a number, plus 27

   $2r + 27$

6. 12 less than 15 times x

   $15x - 12$

7. the product of e and 4, divided by 12

   $4e \div 12$

8. y less than 18 times 6

   $18 \cdot 6 - y$

9. 48 more than the quotient of a number and 64 $m \div 64 + 48$

10. 500 less than the product of 4 and a number $4t - 500$

11. the quotient of p and 4, decreased by 320 $p \div 4 - 320$

12. 13 multiplied by the amount 60 minus w $13(60 - w)$

13. the quotient of 45 and the sum of c and 17 $45 \div (c + 17)$

14. twice the sum of a number and 600 $2(d + 600)$

15. There are twice as many flute players as there are trumpet players in the band. If there are n flute players, write an algebraic expression to find out how many trumpet players there are. $n \div 2$

16. The Nile River is the longest river in the world at 4,160 miles. A group of explorers traveled along the entire Nile in x days. They traveled the same distance each day. Write an algebraic expression to find each day's distance. $4{,}160 \div x$

17. A slice of pizza has 290 calories, and a stalk of celery has 5 calories. Write an algebraic expression to find out how many calories there are in a slices of pizza and b stalks of celery. $290a + 5b$

18. Grant pays 10¢ per minute plus \$5 per month for telephone long distance. Write an algebraic expression for m minutes of long-distance calls in one month. $0.10m + 5$

60
Holt Mathematics

Write each phrase as an algebraic expression.

1. the product of 6 and the square of a number $6n^2$

2. the square of the product of 6 and a number $(6n)^2$

3. 4 times the sum of a number and 6,008 $4(r + 6{,}008)$

4. 200 less than half of a number $y \div 2 - 200$ or $\frac{y}{2} - 200$

5. 3 times the difference of a number squared and 82 $3(n^2 - 82)$

6. 999 less than 45 increased by the product of a number and 85 $(45 + 85e) - 999$

Write a verbal expression for each algebraic expression. Possible answers given.

7. $2(4n)$ twice the product of a number and 4

8. $100 - \dfrac{54}{w}$ 100 less the quotient of 54 and w

9. $r^2 + 4r + 7$ r squared plus 4 times r, plus 7

10. $\dfrac{45}{5s^2}$ 45 divided by the product of 5 and a number squared

11. An albatross can sleep while flying 25 mi/h. An albatross flew 3 miles awake and another n hours asleep at 25 mi/h. Write an algebraic expression to find the distance flown. $25n + 3$

12. You have d dimes, q quarters, and n nickels. Write an algebraic expression to find the total amount of money. $0.10d + 0.25q + 0.05n$

13. Four out of every 10 homes in the United States have a dog. Write an algebraic expression to find out how many dogs there are in h homes. $\dfrac{4h}{10}$

14. A waitress who worked k hours earned \$32 in tips. She gets an additional salary of \$4.50 per hour. Write an algebraic expression to find the amount she earned. $4.50k + 32$

61
Holt Mathematics

Use the operation clues in a word phrase to translate word phrases into algebraic expressions.

Addition		**Subtraction**	
add	plus	subtract	minus
sum	more than	difference	less than
increased by		decreased by	take away

Multiplication	**Division**
times	divided by
multiplied by	divided into
product	quotient

Write an algebraic expression for the difference of a number and 8.

1. What operation would you choose?   subtraction

2. Write an algebraic expression. $n - 8$

Write an algebraic expression for 3 more than a number.

3. What operation would you choose?   addition

4. Write an algebraic expression. $n + 3$

Write an algebraic expression for the quotient of a number and 15.

5. What operation would you choose?   division

6. Write an algebraic expression. $n \div 15$

Write an algebraic expression.

7. the product of 12 and a number k $12k$

8. a number d increased by 9 $d + 9$

9. a number h divided by 4 $h \div 4$

62
Holt Mathematics

Challenge
What's My Equation?

Match each equation with the word problem it represents. Write the equation and corresponding letter. Then write the letter of the equation in the circle that has the problem number. Discover the message formed by the letters.

P $2k = 14$	**D** $\frac{p}{3} = 2.50$	**A** $\frac{m}{5} = 8$	**U** $3t = 21$
M $a - 18 = 33$	**S** $w + 7 = 25$	**T** $x + 7 = 22$	**H** $n - 12 = 37$

1. Tom has 7 more CDs than Rick does. If Tom has 22 CDs, how many does Rick have? $x + 7 = 22;\ T$

2. Maria is 18 years younger than Kim. If Maria is 33, how old is Kim? $a - 18 = 33;\ M$

3. Five friends went out for dinner. They shared the cost of the meal equally. If each person paid $8, what was the total cost of the meal? $\frac{m}{5} = 8;\ A$

4. Max paid 3 times as much for a tape as his friend did. If Max paid $21, how much did his friend pay? $3t = 21;\ U$

5. Marisa and her two friends share a pizza. The cost of the pizza is shared equally among them. If each person pays $2.50, how much does the pizza cost? $\frac{p}{3} = 2.50;\ D$

6. Lee has scored twice as many goals as Jiang has. If Lee's goal total is 14, how many goals has Jiang scored? $2k = 14;\ P$

7. Jamal sold 7 more magazine subscriptions than Wayne did. If Jamal sold 25 subscriptions, how many did Wayne sell? $w + 7 = 25;\ S$

8. Shelly delivered 12 fewer newspapers this week than last week. If she delivered 37 papers this week, how many did she deliver last week? $n - 12 = 37;\ H$

Circles (problem numbers): M 2, A 3, T 1, H 8, A 3, D 5, D 5, S 7, U 4, P 6

63   **Holt Mathematics**

Problem Solving
Translate Words into Math

Write the correct answer.

1. Employers in the United States allocate n fewer vacation days than the 25 days given by the average Japanese employer. Write an algebraic expression to show the number of vacation days given U.S. workers.

   $25 - n$

2. There are 112 members in the Somerset Marching Band. They will march in r equal rows. Write an algebraic expression for the number of band members in each row.

   $112 \div r$

3. A cup of cottage cheese has 26 grams of protein. Write an algebraic expression for the amount of protein in s cups of cottage cheese.

   $26s$

4. Every morning Sasha exercises for 20 minutes. She exercises k minutes every evening. Next week she will double her exercise time at night. Write an algebraic expression to show how long Sasha will exercise each day next week.

   $20 + 2k$

Choose the letter for the best answer.

5. One centimeter equals 0.3937 inches. Which algebraic expression shows how many inches are in c centimeters?
   A $0.3937 + c$
   B $0.3937 \div c$
   C $c \div 0.3937$
   D $0.3937c$

6. In 1957, the Soviet Union launched *Sputnik 1*, the first satellite to orbit Earth. It circled Earth's orbit every 1.6 hours for 92 days, then burned up. If the satellite traveled m miles per hour, which algebraic expression shows the length of the orbit?
   F $92m$
   G $1.6m$
   H $m \div 1.6$
   J $92 \div m$

7. Gina's heart rate is 70 beats per minute. Which algebraic expression shows the number of beats in h hours?
   A $70h$
   B $60h$
   C $4,200h$
   D $3,600h$

8. The Harris family went on vacation for w weeks and 3 days. Which algebraic expression shows the total number of days of their vacation?
   F $7w$
   G $3w$
   H $7w + 3$
   J $3w + 7$

64   **Holt Mathematics**

Reading Strategies
Use a Graphic Organizer

Organizing word phrases for the four operations can help you write algebraic expressions. Study the phrases and the key words highlighted in this visual map.

$r + 12$	$5w$ or $5 \cdot w$
• **add** 12 to a number • a number **plus** 12 • the **sum** of a number and 12 • 12 **more than** a number • a number **increased by** 12	• 5 **times** a number • 5 **multiplied by** a number • the **product** of 5 and a number

Word Phrases for Algebraic Expressions

$n - 8$	$r \div 4$ or $\frac{r}{4}$
• **subtract** 8 from a number • a number **minus** 8 • 8 **less than** a number • a number **decreased by** 8 • **take away** 8 from a number	• 4 **divided into** a number • the **quotient** of a number with a divisor of 4 • a number **divided by** 4

Write a word phrase for each algebraic expression.

1. $n - 15$ possible answer: 15 less than n

2. $(m + 12) - 3$ possible answer: 3 less than the sum of m and 12

3. $2(y + 8)$ possible answer: twice the sum of y and 8

4. $5 + \frac{w}{3}$ possible answer: 5 increased by the quotient of a number and 3

Write an algebraic expression for each word phrase.

5. a number decreased by 9 $n - 9$

6. the product of 15 and r $15r$

7. the quotient of a number with a divisor of 5 $p \div 5$

8. three times the sum of n and 6 $3(n + 6)$

65   **Holt Mathematics**

Puzzles, Twisters & Teasers
Birds of a Feather!

Solve the crossword puzzle. Then use the letters in the shaded boxes to answer the riddle. You'll need to use some letters more than once.

Across

1. something that does not change
2. a number or symbol placed to the right of and above another number, symbol, or expression
6. putting together
8. the letter in a term
9. a number, a variable, or a product of numbers and variables

Down

1. the number in a term
3. the multiple expressed by an exponent
4. a symbol used for counting
5. group similar objects
7. similar

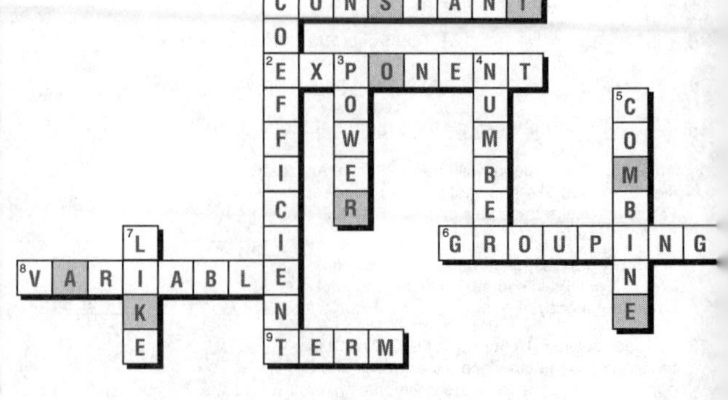

Where do birds invest their money?

In the S T O R K   M A R K E T .

66   **Holt Mathematics**

Practice A
Simplifying Algebraic Expressions

Identify like terms in each list.

1. $6a$ b a 17 $4b$ 32 $17a$

   $6a$, a, and $17a$; b and $4b$; 17 and 32

2. x x^2 $3x$ 3 $3x^2$ 6

   x^2 and $3x^2$; x and $3x$; 3 and 6

3. 2 $6z$ $6z^2$ z $17z$ z^2 3

   $6z^2$ and z^2; z, $6z$, and $17z$; 2 and 3

4. m 8 $8m^2$ $8m$ m^2 $12m$ 18

   $8m^2$ and m^2; m, $8m$, and $12m$; 8 and 18

5. $2p$ $22p$ $56q$ 12^2 q 34

   $2p$ and $22p$; $56q$ and q; 12^2 and 34

6. d d^2 $15d^2$ $2d$ 4^2 $5d$ 44

   d, $2d$, and $5d$; d^2 and $15d^2$; 4^2 and 44

Combine like terms.

7. $6p^2 + 3p^2$

   $9p^2$

8. $9x - 6x$

   $3x$

9. $a^2 + b^2 + 2a^2 + 5b^2$

   $3a^2 + 6b^2$

10. $7h^2 + 3 - 2h^2 + 4$

   $5h^2 + 7$

11. $3x + 3y + x + y + z$

   $4x + 4y + z$

12. $5b + 5b + 6b^2 - 10 - 3b$

   $6b^2 + 7b - 10$

13. Find the perimeter of the rectangle. Combine like terms.

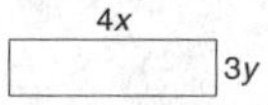

   A $4x + 3y$
   B $8x + 6y$
   C $12xy$
   D $4x^2 + 3y^2$

(rectangle labeled $4x$ by $3y$)

67

Practice B
Simplifying Algebraic Expressions

Identify like terms in each list.

1. $3a$ b^2 b^3 $4b^2$ 4 $5a$

   $3a$ and $5a$; b^2 and $4b^2$

2. x x^4 $4x$ $4x^2$ $4x^4$ $3x^2$

   x and $4x$; x^4 and $4x^4$; $4x^2$ and $3x^2$

3. $6m$ $6m^2$ n^2 $2n$ 2 $4m$ $5n$

   $6m$ and $4m$; $2n$ and $5n$

4. $12s$ $7s^4$ $9s$ s^2 5 $5s^4$ 2

   $12s$ and $9s$; $7s^4$ and $5s^4$; 5 and 2

Simplify. Justify your steps using the Commutative, Associative, and Distributive Properties when necessary.

5. $2p + 22q^2 - p$

   $p + 22q^2$

6. $x^2 + 3x^2 - 4^2$

   $4x^2 - 16$

7. $n^4 + n^3 + 3n - n - n^3$

   $n^4 + 2n$

8. $4a + 4b + 2 - 2a + 5b - 1$

   $2a + 9b + 1$

9. $32m^2 + 14n^2 - 12m^2 + 5n - 3$

   $20m^2 + 14n^2 + 5n - 3$

10. $2h^2 + 3g - 2h^2 + 2^2 - 3 + 4g$

   $7g + 1$

11. Write an expression for the perimeter of the figure at the right. Then simplify the expression.

   $2v + 8s + 5$

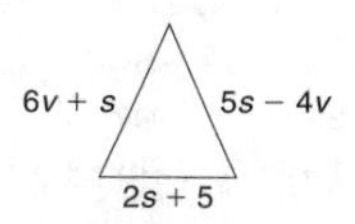

12. Write an expression for the combined perimeters of the figures at the right. Then simplify the expression.

   $14a + 2b + 4$

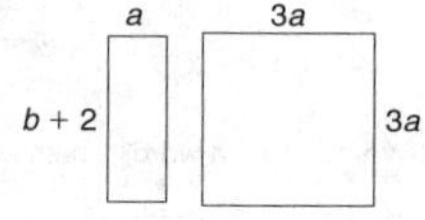

68

Practice C
Simplifying Algebraic Expressions

Simplify each expression. Justify your steps using the Commutative, Associative, and Distributive Properties, when necessary.

1. $8k^2 + 4k - 3k^2 + 3^2 - k + 5$

   $5k^2 + 3k + 14$

2. $10x^3 + 5y^2 + 2xy - 4y^2 + 4xy - x^3$

   $9x^3 + y^2 + 6xy$

3. $3a + 2b^2 + 6c + a - 2c + b^2 + c$

   $4a + 3b^2 + 5c$

4. $12x^4 + 6x^2 + 5x^3 - x^2 + 2xy - 8x^4$

   $4x^4 + 5x^3 + 5x^2 + 2xy$

5. $9p^6 + q^2 + 6p + 5q^2 + 5p - 5q^2$

   $9p^6 + q^2 + 11p$

6. $h^2 + 4h + 4h^2 - h + 4 + h^2 + 7h$

   $6h^2 + 10h + 4$

7. Write an expression that has five terms and simplifies to $5m^3 + 4n$.

   Possible answer: $3m^3 + 5n + 4m^3 - 2m^3 - n$

8. Write and simplify an expression for the perimeter of the figure to the right.

   $6x + 7y + 5$

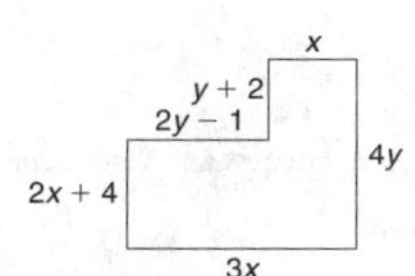

9. Write an expression to find the combined perimeters of the figures to the right. Then simplify the expression.

   $12a + 14b + 12$

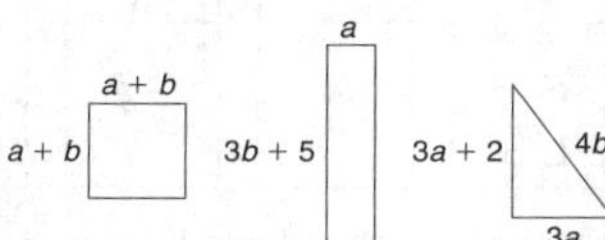

10. Jake scored x points in the first basketball game. He scored 2 fewer points in the next game. His teammate, Jack, scored $2y$ points in the first game and 4 more than twice as many points in the next game. Write and simplify an expression for the total number of points scored by both players.

   $2x + 6y + 2$

69

Reteach
Simplifying Algebraic Expressions

Look at the following expressions:
$$x = 1x$$
$$x + x = 2x$$
$$x + x + x = 3x$$
The numbers 1, 2, and 3 are called **coefficients** of x.

Identify each coefficient.

1. $3n$ 3 2. $7y$ 7 3. m 1 4. 9 0

An algebraic expression has terms that are separated by $+$ and $-$. In the expression $2x + 5y$, the **terms** are $2x$ and $5y$.

Expression	Terms
$8x + 4y$	$8x$ and $4y$
$m - 3n$	m and $3n$
$4a^2 - 2b + a$	$4a^2$, $2b$, and a
$6d + 2p$	$6d$ and $2p$

Sometimes the terms of an expression can be combined. Only **like terms** can be combined.

$7w + w$ like terms

$2x - 2y + 2$ unlike terms because x and y are different variables

$8e - 3e + 2e$ like terms

$5d + 25g$ unlike terms because d and g are different variables

To simplify an expression:
Step 1: Combine like terms.
Step 2: Add or subtract the coefficients of the variable.

$$7w + w = 8w$$
$$6y + 1 - 3y = 3y + 1$$

Simplify.

5. $y + 5y$

   $6y$

6. $9x - 4x$

   $5x$

7. $5s - 2s$

   $3s$

8. $3d + 7d$

   $10d$

9. $3b + b + 6$

   $4b + 6$

10. $8a - a - 3$

   $7a - 3$

11. $2p + 4p + r$

   $6p + r$

12. $9b - 8b + c$

   $b + c$

70

115

Challenge
Matching Terms

Draw a line from each set of terms in Column A to its equivalent combination in Column B. Then circle each letter in Column B that does not have a matching term. Unscramble those letters to answer the riddle.

Column A

1. $2x + 7 + 5x - 4 - x$
2. $5 + 7x + 2x - 3 + 6$
3. $x + y + 4x - 3x + 2y + 3y$
4. $3x^2 + 5x - 17 + 6x + 20$
5. $4x + x^2 + 12 - 4 + 2x$
6. $12y + 12x + 12 - 6x + 12$
7. $12y + 4 + x - 7y + 8 + 8x$
8. $5x + x^2 + 2x + 5 - 4 - x^2$
9. $5x^2 + 8x + 7x^2 + 6x$
10. $12x + 6 - 8x - 4x - 3 + 12$
11. $5x + 4 - 3x + 5 + 2x - 9$
12. $4x + 2y + 8 - 3 - y - x$
13. $4x + 5 + 7x + 2y + 2 - y$
14. $2y + 2x + 8 - 6 + x - 2y$
15. $4x + 6y + 6 + 7x + y$
16. $3x^2 + 4x - 2x^2 - 3x + 2x$
17. $8x + 4 - 4 - 4x + x$
18. $y + 5x + 6y + 9 - 6$
19. $x^2 + 3 + 2x^2 + 4 - 7$
20. $5y + 3 + 7x^2 - 2 - x^2 + y$

Column B

A. $5y + 9x + 12$
B. $12y + 6x + 24$
C. 15
D. $9x + 8$
E. 4
F. $6x + 3$
G. $11x + y + 7$
H. $x^2 + 6x + 8$
I. $4x$
J. $3x^2 + 11x + 3$
K. $3x + 2$
L. $3x^2$
M. $6x$
N. $x^2 + 3x$
O. $6x^2 + 6y + 1$
P. $12x^2 + 14x$
Q. $7x + 1$
R. x^2
S. $5x + 7y + 3$
T. 0
U. $2x + 6y$
V. $3x + y + 5$
W. $11x + 7y + 6$
X. $5x$

Riddle: What can be a word, a number, a period of time, or a variable?

A <u>T</u> <u>E</u> <u>R</u> <u>M</u>

71   Holt Mathematics

Problem Solving
Simplifying Algebraic Expressions

Write the correct answer. Use the figures for Problems 1–3.

1. Figure 1 shows the length of each side of a garden. Write and simplify an expression for the perimeter of the garden.

   $2a + 2b + 2c$

2. Figure 2 is a square swimming pool. Write and simplify an expression for the perimeter of the pool.

   $4b$

3. Write and simplify an expression for the combined perimeter of the garden and the pool.

   $2a + 6b + 2c$

Figure 1   Figure 2

4. The Pantheon in Rome has n granite columns in each of 3 rows. Write and simplify an addition expression to show the number of columns. Then evaluate the expression for $n = 8$.

   $n + n + n; 3n = 24;$

   24 columns

Choose the letter for the best answer.

5. Which is an expression that shows the earnings of a telemarketer who worked for 23 hours at a salary of d dollars per hour?

   A $d + 23$ C $d \div 23$
   (B) $23d$ D $23 \div d$

6. The minimum wage set in 1997 was $5.15 per hour. Evaluate the expression $40h$ where $h = \$5.15$ to find a worker's weekly salary.

   F $20.60   H $515.00
   G $200   (J) $206.00

7. What is the perimeter of a triangle with sides the following lengths: $2a + 4c$, $3c + 7$, and $6a - 4$. Simplify the expression.

   A $8a + 11c$
   B $6a + 7c + 3$
   (C) $8a + 7c + 3$
   D $8a + 7c + 11$

8. A hexagon is a 6-sided figure. Find the perimeter of a hexagon where all of the sides are the same length and the expression $x + y$ represents the length of a side. Simplify the expression.

   (F) $6x + 6y$
   G $6 + x + y$
   H $6x + y$
   J $6xy$

72   Holt Mathematics

Reading Strategies
Organization Patterns

An algebraic expression is made up of parts called **terms**.

constants	variables	constants and variables
$3.2 \quad \frac{1}{2} \quad 12$	$m \quad s \quad x$	$\frac{n}{2} \quad \frac{2}{3}y \quad 4x \quad 3m^2$

A **coefficient** is a value multiplied by a variable.

Term	Value of Coefficient	Meaning
$7x$	7	$7 \cdot x$
y	1	$1 \cdot y$
$\frac{x}{3}$	$\frac{1}{3}$	$\frac{1}{3} \cdot x$

This expression has 6 terms:

Term   Term   Term   Term   Term   Term

$2x \; + \; 5b \; + \; 7 \; - \; b \; + \; 3x \; + \; 2x^2$

- Reorganize the terms: $2x + 3x + 5b - b + 7 + 2x^2$.
- Combine like terms: $5x + 4b + 7 + 2x^2$.

Answer each question.

1. How many terms are there in this expression:
   $6b + b^2 + 5 + 2b - 3f$?

   5 terms

2. $6b$ and b^2 are unlike terms. Explain why.

   In b^2, b is raised to the second power, but it is not in $6b$.

3. How many terms are there in this expression:
   $5a^2 + 6b + a^3 + 2a^2 - 3b - 2$?

   6

4. Reorganize these terms so like terms are next to each other.

   $a^3 + 5a^2 + 2a^2 + 6b - 3b - 2$

73   Holt Mathematics

Puzzles, Twisters & Teasers
In Other Words...

Write each verbal expression as an algebraic expression. Then use the answer key to solve the riddle.

1. the product of 20 and t $20t$

2. the sum of 4 times a number and 2 $4n + 2$

3. the product of 7 and p $7p$

4. the sum of six times a number and 1 $6n + 1$

5. the sum of 5 and a number $5 + n$

6. the quotient of a number and 8 $\frac{n}{8}$

7. m plus 6 $m + 6$

8. t less than 23 $23 - t$

9. the quotient of 100 and the amount 6 plus w $\frac{100}{(6 + w)}$

Answer Key

+	−	×	÷
S H L B	N	G A	C I

What's worse than raining cats and dogs?

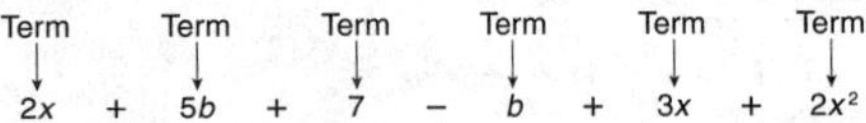

<u>H</u> <u>A</u> <u>I</u> <u>L</u> <u>I</u> <u>N</u> <u>G</u>
+   ×   ÷   +   ÷   −   ×

<u>C</u> <u>A</u> <u>B</u> <u>S</u>
÷   ×   +   +

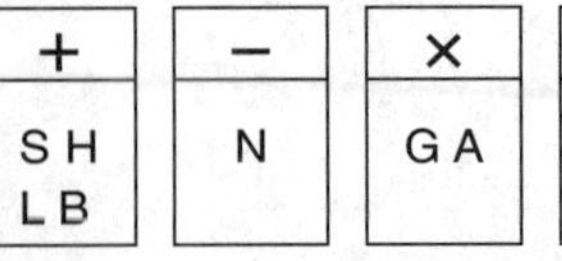

74   Holt Mathematics

Holt Mathematics

Practice A
Equations and Their Solutions

Tell if the value of the variable is a solution for the equation.

1. $n = 22$ for $n + 5 = 27$ **2.** $a = 43$ for $72 - a = 30$ **3.** $g = 31$ for $19 + g = 40$

    yes   no   no

4. $z = 87$ for $z - 22 = 65$ **5.** $f = 15$ for $46 + f = 61$ **6.** $h = 53$ for $h - 24 = 77$

    yes   yes   no

7. $p = 4$ for $5p = 20$ **8.** $k = 48$ for $k \div 8 = 8$ **9.** $m = 12$ for $48 \div m = 4$

    yes   no   yes

10. $x = 8$ for $32x = 264$ **11.** $v = 5$ for $\frac{90}{v} = 15$ **12.** $s = 7$ for $16s = 112$

    no   no   yes

13. $m = 2$ for $3m + 4 = 10$ **14.** $y = 6$ for $3y - 4 = 18$ **15.** $r = 10$ for $10r - 10 = 90$

    yes   no   yes

16. In the United States in 2005, there were 68 endangered species that were mammals. There were 9 more endangered species that were birds. Were there 59 or 77 endangered species of birds in the United States?

77 endangered species

17. At the Bike and Blade Shop, mountain bikes are on sale for $349. This is $30 more than a racing bike costs. Does the racing bike cost $319 or $379?

$319

18. Which problem situation best matches the equation $2c + 10 = 350$?

Situation A: Austin paid $10 for a new video game console. This is $350 more than 2 times the cost of a used console. How much does a used console cost?

Situation B: Austin paid $350 for a new video game console. This is $10 more than 2 times the cost of a used console. How much does a used console cost?

Situation B

75   **Holt Mathematics**

Practice B
Equations and Their Solutions

Determine whether the given value of the variable is a solution.

1. $a = 4$ for $12 - a = 6$ **2.** $m = 37$ for $23 + m = 60$ **3.** $x = 6$ for $54 = 9x$

    no   yes   yes

4. $g = 96$ for $\frac{g}{4} = 32$ **5.** $n = 28$ for $n + 44 = 72$ **6.** $j = 6$ for $84 \div j = 12$

    no   yes   no

7. $k = 24$ for $3k = 6$ **8.** $m = 3$ for $42 = m + 39$ **9.** $y = 8$ for $8y + 6 = 70$

    no   yes   yes

10. $s = 5$ for $18 = 3s - 3$ **11.** $k = 7$ for $23 - k = 30$ **12.** $v = 12$ for $84 = 7v$

    no   no   yes

13. $c = 15$ for $45 - 2c = 15$ **14.** $x = 10$ for $x + 25 - 2x + 4 = 19$

    yes   yes

15. $e = 6$ for $42 = 51 - e$ **16.** $p = 15$ for $19 = p - 4$

    no   no

17. Jason and Maya have their own web sites on the Internet. As of last week, Jason's web site had 2,426 visitors. This is twice as many visitors as Maya had. Did Maya have 1,213 visitors or 4,852 visitors to her web site?

1,213 visitors

18. Which problem situation best matches the equation $3c - 5 = 31$?

Situation A: Rachel had a coupon for $5 off the cost of her order. She ordered 3 large pizzas that each cost the same amount and paid a total of $31. What was the cost of each pizza?

Situation B: Rachel had a coupon for $31 off the cost of her order. She ordered 5 large pizzas that each cost the same amount and paid a total of $3. What was the cost of each pizza?

Situation A

76   **Holt Mathematics**

Practice C
Equations and Their Solutions

Determine whether the given value of the variable is a solution.

1. $a = 15$ for $75 \div a = 5$ **2.** $x = 90$ for $x \div 9 = 100 - x$

    yes   yes

3. $d = 8$ for $875 = 909 - 4d$ **4.** $x = 32$ for $2x - 25 + x - 70 = 1$

    no   yes

5. $e = 2$ for $e^3 - e^2 = 6e - 8$ **6.** $b = 4$ for $b^2 + 2b - 3 = 27$

    yes   no

7. $d = 12$ for $4d - 24 - 12 = 0$ **8.** $r = 9$ for $4r^2 - 19 - 5r = 340$

    no   no

9. $p = 25$ for $\frac{4p}{5} + 2p + 10 = 80$ **10.** $t = 18$ for $\frac{t}{6} + \frac{54}{t} = 3$

    yes   no

11. In 1993, there were about 34,000,000 cell phone subscribers in the United States. This is about 125,000,000 fewer subscribers than there were by the year 2003. The equation $34,000,000 = s - 125,000,000$ can be used to represent the number of cell phone subscribers in 2003. Were there 91,000,000 or 159,000,000 cell phone subscribers in the United States in 2003?

159,000,000 cell phone subscribers

12. Which problem situation best matches the equation $50n + 15 = 150$?

Situation A: Hector paid a commission of $150 to buy 15 shares of an Internet stock. Altogether he paid $50. How much did each share of stock cost?

Situation B: Hector paid a commission of $15 to buy 50 shares of an Internet stock. Altogether he paid $150. How much did each share of stock cost?

Situation B

77   **Holt Mathematics**

Reteach
Equations and Their Solutions

Number sentences that contain an equal sign (=) are called **equations**.

Equations may be true, or they may be false.

     True   False
     $3 + 4 = 7$ $3 + 1 = 7$
     $8 - 6 = 2$ $8 - 2 = 5$

An equation may contain a variable.

     variable → $x + 4 = 6$

Whether this equation is true or false depends on the value of x.

You can decide if a number is a *solution* of an equation. Substitute the number for the variable in the equation. If the equation is a true equation, then the number is the **solution**.

Equation: $x + 4 = 6$

Is 2 a solution?   Is 3 a solution?

     $x + 4 = 6$ $x + 4 = 6$

Substitute 2 for x Substitute 3 for x

     $2 + 4 \stackrel{?}{=} 6$ $3 + 4 \stackrel{?}{=} 6$
     $6 \stackrel{?}{=} 6$ $7 \stackrel{?}{=} 6$

2 is a solution of $x + 4 = 6$.   3 is not a solution of $x + 4 = 6$.

Tell if the number is a solution.

1. Is 3 a solution of $y + 3 = 9$?   **2.** Is 4 a solution of $n + 6 = 10$?

    no   yes

3. Is 2 a solution of $w - 1 = 1$?   **4.** Is 1 a solution of $x + 50 = 49$?

    yes   no

5. Is 6 a solution of $c + 23 = 30$?   **6.** Is 9 a solution of $v - 9 = 0$?

    no   yes

7. Is 20 a solution of $t - 17 = 3$?   **8.** Is 16 a solution of $12 + a = 24$?

    yes   no

9. Is 25 a solution of $38 - m = 13$?   **10.** Is 8 a solution of $15 = e + 5$?

    yes   no

78   **Holt Mathematics**

Challenge
The Solution Is BINGO!

Find the solution to each problem or equation. Cross it out on the board below to get BINGO!

1. Is 37, 47, or 67 a solution for $52 = n + 15$?

__37__

2. Is 14, 17, or 21 a solution for $8y - 7 = 129$?

__17__

3. Is 14, 22, or 24 a solution for $132 - (4x - 5) = 81$?

__14__

4. Is 12, 15, or 18 a solution for $3(60 - s) - 2s = 105$?

__15__

5. Garret scored 18 points in his last basketball game, which is 6 fewer points than Vince scored. The equation $18 = p - 6$ can be used to represent Vince's points. Did Vince score 12, 24, or 28 points?

__24__

6. In 3 years, Sarah's sister will be twice as old as Sarah. If Sarah is now 3 years old, will her sister be 6, 9, or 12 years old in 3 years?

__12__

7. The highest recorded temperature in Alaska was 100°F in 1915. This was 56 years before the lowest recorded temperature of −80°F. Was the lowest recorded temperature in 1856, 1956, or 1971?

__1971__

8. In 2002, Florida had 3,314 public schools. This was 155 more public schools than 3 times the number of public schools in South Carolina during the same year. Did South Carolina have 1,053, 2,849, or 3,159 public schools in 2002?

__1,053__

B	I	N	G	O
17	67	24	37	1971
6	24	47	3,159	22
2,849	1956	FREE	18	14
12	9	16	1,053	1856
28	23	21	19	15

79
Holt Mathematics

Problem Solving
Equations and Their Solutions

Write the correct answer.

1. The jet airplane was invented in 1939. This is 12 years after the first television was invented. Was television invented in 1927 or 1951?

__1927__

2. There are three times as many students in the high school as in the junior high school, which has 330 students. Does the high school have 990 students or 110 students?

__990 students__

3. The frigate bird has been recorded at speeds up to 95 mi/h. The only faster bird ever recorded was the spine-tailed swift at 11 mi/h faster. Was the speed of the spine-tailed swift 84 mi/h or 106 mi/h?

__106 mi/h__

4. As of 2004, there were 20.5 million Internet users in Canada. This is 6.6 million more Internet users than there were in Mexico. Were there 27.1 million or 13.9 million Internet users in Mexico?

__13.9 million Internet users__

Choose the letter for the best answer.

5. In the United States, the average school year is 180 days. This is 71 days less than the average school year in China. What is the average school year in China?

A) 251 days
B 109 days
C 151 days
D 271 days

6. The longest suspension bridge in the world is the Akashi Kaikyo Bridge in Japan. Its main span is 1,290 feet longer than a mile. A mile is 5,280 feet. How long is the Akashi Kaikyo bridge?

F 3,990 feet
G 6,400 feet
H 4,049 feet
J) 6,570 feet

7. *Ornithomimus* stood about 6 feet tall and was the fastest dinosaur at a speed of about 50 mi/h. The largest dinosaur, *Seismosaurus*, was 20 times as tall. How tall was *Seismosaurus*?

A 12 feet
B 70 feet
C) 120 feet
D 26 feet

8. Milton collects sports trading cards. He has 80 baseball cards. He has half as many basketball cards as football cards. He has 20 more hockey cards than basketball cards and half as many football cards as baseball cards. How many hockey cards does he have?

F 20 hockey cards
G) 40 hockey cards
H 60 hockey cards
J 80 hockey cards

80
Holt Mathematics

Reading Strategies
Use a Visual Model

The expressions on both sides of the equal sign are equal in an **equation**.

You can read an equation in math much like you read a sentence. When you see an equal sign, you read, **"is the same as."** Balanced scales can help you picture an equation.

Mark has 45 baseball cards. This is 8 more than his sister Kathy. → Mark's cards are equal to Kathy's cards + 8.

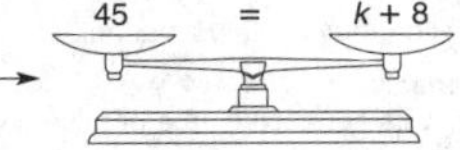

$$45 = k + 8$$

The balanced scales illustrate this equation. →

The value of the variable that makes the equation a true sentence is called the **solution**.

Substitute 40 for k: $45 = k + 8$
$45 \neq 40 + 8$
40 is not a solution.

Substitute 37 for k: $45 = k + 8$
$45 = 37 + 8$
37 is the solution.

Answer each question.

1. What is an equation?

__a mathematical sentence with expressions of the same value on both sides of the equal sign__

2. What is the solution of an equation?

__sample answer: the value of the variable that makes the expressions equal__

3. Is $3 + 28 \neq 32$ an equation? Why or why not.

__no; The expressions are not equal in value.__

4. Is $n = 12$ a solution for the equation $43 = n + 31$? Why or why not?

__yes; $43 = 12 + 31$__

Determine whether the given value of the variable is a solution.

5. $d + 37 = 54$ for $d = 23$ ____ $d = 23$ is not a solution.

6. $t - 23 = 42$ for $t = 56$ ____ $t = 56$ is not a solution.

81
Holt Mathematics

Puzzles, Twisters & Teasers
The Answer Popped Right into My Head!

Find the solution for each equation below. Write the letter on the line above the correct answer at the bottom of the page to solve the riddle.

P $18 = s - 7$ __25__
M $x + 3 = 10$ __7__
O $12 = t + 9$ __3__
Y $s - 38 = 57$ __95__
R $16 - j = 12$ __4__
I $16 = 34 - m$ __18__
U $48 = x + 12$ __36__
S $17 + k = 40$ __23__
N $24 = 34 - n$ __10__
B $p + 18 = 29$ __11__
A $47 = v - 6$ __53__
G $94 = c + 6$ __88__
H $82 = j + 9$ __73__
E $a - 15 = 17$ __32__
T $15 = k - 9$ __24__

What did one firecracker say to the other?

M	Y		P	O	P		I	S
7	95		25	3	25		18	23

B	I	G	G	E	R		T	H	A	N
11	18	88	88	32	4		24	73	53	10

Y	O	U	R		P	O	P
95	3	36	4		25	3	25

82
Holt Mathematics

Match each equation in Column A with its correct solution in Column B.

Column A	Column B		Column A	Column B
1. $n - 16 = 8$	A. $n = 12$		10. $x - 12 = 13$	L. $x = 14$
2. $5 = n - 7$	B. $n = 13$		11. $x + 8 = 40$	M. $x = 17$
3. $12 + n = 25$	C. $n = 17$		12. $34 = 16 + x$	N. $x = 18$
4. $n - 17 = 11$	D. $n = 24$		13. $x + 5 = 19$	P. $x = 25$
5. $n + 18 = 35$	E. $n = 27$		14. $4 + x = 52$	Q. $x = 32$
6. $7 = n - 28$	F. $n = 28$		15. $12 + x = 50$	R. $x = 33$
7. $n - 12 = 40$	G. $n = 35$		16. $15 = x - 2$	S. $x = 38$
8. $24 = n - 25$	H. $n = 49$		17. $52 = x + 9$	T. $x = 43$
9. $46 = n + 19$	J. $n = 52$		18. $x - 11 = 22$	U. $x = 48$

19. Chris has 55 baseball trading cards. He has 17 more cards than his sister Sara has. Write and solve an equation to find how many trading cards Sara has.

$55 = 17 + n; n = 38$

20. In 2000, Sammy Sosa hit 50 home runs. His home run total was 23 runs fewer than the number of home runs that Barry Bonds hit the next year. Write and solve an equation to find how many home runs Barry Bonds hit in 2001.

$50 = h - 23; h = 73$

Holt Mathematics

Solve each equation. Check your answer.

1. $33 = y - 44$

$y = 77$

2. $r - 32 = 77$

$r = 109$

3. $125 = x - 29$

$x = 154$

4. $k + 18 = 25$

$k = 7$

5. $589 + x = 700$

$x = 111$

6. $96 = 56 + t$

$t = 40$

7. $a - 9 = 57$

$a = 66$

8. $b - 49 = 254$

$b = 303$

9. $987 = f - 11$

$f = 998$

10. $32 + d = 1,400$

$d = 1,368$

11. $w - 24 = 90$

$w = 114$

12. $95 = g - 340$

$g = 435$

13. $e - 35 = 59$

$e = 94$

14. $84 = v + 30$

$v = 54$

15. $h + 15 = 81$

$h = 66$

16. $110 = a + 25$

$a = 85$

17. $45 + c = 91$

$c = 46$

18. $p - 29 = 78$

$p = 107$

19. $56 - r = 8$

$r = 48$

20. $39 = z + 8$

$z = 31$

21. $93 + g = 117$

$g = 24$

22. The Morales family is driving from Philadelphia to Boston. So far, they have driven 167 miles. This is 129 miles less than the total distance they must travel. How many miles is Philadelphia from Boston?

The total distance is 296 miles.

23. Ron has $1,230 in his savings account. This is $400 more than he needs to buy a new big screen TV. Write and solve an equation to find out how much the TV costs.

$1,230 = t + 400;$ The TV costs $830.

Holt Mathematics

Solve each equation. Check your answer.

1. $b - 32 = 15$

$b = 47$

2. $e - 43 = 121$

$e = 164$

3. $601 = x - 24$

$x = 625$

4. $m + 45 = 123$

$m = 78$

5. $314 + z = 350$

$z = 36$

6. $840 = 45 + f$

$f = 795$

7. $d - 67 = 23$

$d = 90$

8. $w + 233 = 319$

$w = 86$

9. $91 = x + 52$

$x = 39$

10. $150 + y = 879$

$y = 729$

11. $k - 32 = 217$

$x = 249$

12. $408 = s - 129$

$s = 537$

13. $108 = j - 24$

$j = 132$

14. $1,204 = w + 389$

$w = 815$

15. $p - 167 = 321$

$p = 488$

16. In 2003, *USA Today* was the leading U.S. daily newspaper, with a circulation of 2,154,539. This was 63,477 more than the second leading daily newspaper, the *Wall Street Journal*. Write and solve an equation to find the circulation of the *Wall Street Journal* in 2003.

$2,154,539 = c + 63,477; c = 2,091,062$

17. Mrs. Baker's class has been raising money for a local charity. At the end of last week, they had collected $238. By the end of this week, they had a total of $419. Write and solve an equation to find the amount collected this week.

$238 + m = 419; m = 181$

18. Andy Green broke the sound barrier on land on October 15, 1997, in Black Rock Desert, Nevada. The speed of sound was recorded at about 751 mi/h. This was about 12 mi/h less than the land speed record that he set that day. Write and solve an equation to find Andy Green's land speed record.

$751 = r - 12; r = 763$ mi/h

Holt Mathematics

Solving an equation is like balancing a scale. If you add the same weight to both sides of a balanced scale, the scale will remain balanced. You can use this same idea to solve an equation.

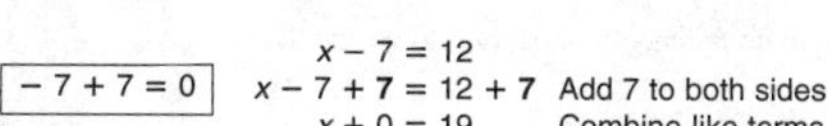

Think of the equation $x - 7 = 12$ as a balanced scale. The equal sign keeps the balance.

$$x - 7 = 12$$
$$\boxed{-7 + 7 = 0} \quad x - 7 + 7 = 12 + 7 \quad \text{Add 7 to both sides.}$$
$$x + 0 = 19 \quad \text{Combine like terms.}$$
$$x = 19$$

When you solve an equation, the idea is to get the variable by itself. What you do to one side of the equation, you must do to the other side.

• To solve a subtraction equation, use addition.
• To solve an addition equation, use subtraction.

Solve and check: $y + 8 = 14$.

$$y + 8 = 14$$
$$\boxed{+8 - 8 = 0} \quad y + 8 - 8 = 14 - 8 \quad \text{Subtract 8 from both sides.}$$
$$y + 0 = 6 \quad \text{Combine like terms.}$$
$$y = 6$$

Check: $y + 8 = 14$ To check, substitute 6 for y.
$$6 + 8 \stackrel{?}{=} 14$$
$$14 \stackrel{?}{=} 14 \quad ✔$$

A true sentence, $14 = 14$, means the solution is correct.

Solve and check.

1. $x - 2 = 8$

$x - 2 + 2 = 8 + 2$

$x - 0 = 10$

2. $b + 5 = 11$

$b + 5 - 5 = 11 - 5$

$b + 0 = 6$

3. $n + 8 = 11$

$n = 3$

4. $y - 6 = 2$

$y = 8$

5. $a - 9 = 4$

$a = 13$

6. $m + 2 = 18$

$m = 16$

Holt Mathematics

119

Holt Mathematics

Challenge
Equation Maker

Use each term once to make up one addition and one subtraction equation, then solve the equations. **Possible answers given.**

1. m, n, 12, 6, 54, 9

$n - 12 = 54; n = 66$

$m + 6 = 9; m = 3$

2. x, y, 7, 15, 32, 45

$x - 15 = 32; x = 47$

$y + 7 = 45; y = 38$

3. p, q, 19, 44, 72, 8

$p + 19 = 72; p = 53$

$q - 8 = 44; q = 52$

4. a, b, 67, 102, 6, 8

$a - 102 = 6; a = 108$

$b + 8 = 67; b = 59$

5. c, d, 11, 12, 18, 35

$c - 18 = 35; c = 53$

$d + 11 = 12; d = 1$

6. s, t, 115, 123, 32, 0

$s + 115 = 123; s = 8$

$t - 32 = 0; t = 32$

7. w, y, 1, 2, 3, 4

$w - 2 = 1; w = 3$

$y + 3 = 4; y = 1$

8. n, p, 6, 22, 99, 400

$n - 6 = 400; n = 406$

$p + 22 = 99; p = 77$

9. e, f, 52, 4, 75, 18

$e + 4 = 75; e = 71$

$f - 18 = 52; f = 70$

10. g, h, 61, 88, 94, 117

$g - 117 = 61; g = 178$

$h + 88 = 94; h = 6$

11. k, l, 302, 54, 115, 79

$k - 54 = 115; k = 169$

$l + 79 = 302; l = 223$

12. r, s, 90, 14, 71, 15

$r + 15 = 90; r = 75$

$s - 14 = 71; s = 85$

13. u, v, 8, 12, 37, 44

$u - 37 = 44; u = 81$

$v + 8 = 12; v = 4$

14. x, y, 198, 0, 231, 4

$x - 198 = 4; x = 202$

$y + 0 = 231; y = 231$

Holt Mathematics

Problem Solving
Solving Equations by Adding or Subtracting

Write the correct answer.

1. In an online poll, 1,927 people voted for Coach as the best job at the Super Bowl. The job of Announcer received 8,055 more votes. Write and solve an equation to find how many votes the job of Announcer received.

$1,927 = v - 8,055; v = 9,982;$

9,982 votes

2. In 2005, the largest bank in the world was UBS, Switzerland, with $1,533 billion in assets. This was $49 billion more than the largest bank in the United States, Citigroup. Write and solve an equation to find Citigroup's assets.

$1,533 = a + 49; a = 1,484;$

$1,484 billion

3. The two smallest countries in the world are Vatican City and Monaco. Vatican City is 1.37 square kilometers smaller than Monaco, which is 1.81 square kilometers in area. What is the area of Vatican City?

0.44 square km

4. The Library of Congress is the largest library in the world. It has 29 million books, which is 10 million more than the National Library of Canada has. How many books does the National Library of Canada have?

19 million books

Choose the letter for the best answer.

5. The first track on Sean's new CD has been playing for 55 seconds. This is 42 seconds less than the time of the entire first track. How long is the first track on this CD?

A 37 seconds   C 97 seconds

B 63 seconds   D 93 seconds

6. There are 45 students on the school football team. This is 13 more than the number of students on the basketball team. How many students are on the basketball team?

F 58 students   H 32 students

G 48 students   J 42 students

7. A used mountain bike costs $79.95. This is $120 less than the cost of a new one. If c is the cost of the new bike, which equation can you use to find the cost of a new bike?

A $79.95 = c + 120$

B $120 = 79.95 - c$

C $79.95 = c - 120$

D $120 = 79.95 + c$

8. The goal of the School Bake Sale is to raise $125 more than last year's sale. Last year the Bake Sale raised $320. If it reaches its goal, how much will the Bake Sale raise this year?

F $445

G $195

H $525

J $425

Holt Mathematics

Reading Strategies
Follow a Procedure

In order to solve an equation, you must find the **solution**. The solution is the value that makes the equation true. To solve an equation, you need to get the variable by itself on one side of the equal sign.

- If you have an addition equation, you must subtract to get the variable by itself.

- If you have a subtraction equation, you must add to get the variable by itself.

Example:

$z + 12 = 32$ ← To get z by itself, subtract 12.

$z + 12 - 12 = 32 - 12$ ← Rewrite the equation to show that 12 is subtracted from both sides.

$z = 20$ ← This is the solution after subtracting 12 from both sides.

Check by using 12 in place of z.

$20 + 12 \stackrel{?}{=} 32$

$32 = 32$, so $z = 20$ is the correct solution.

Example:

$27 = x - 8$ ← To get x by itself, add 8.

$27 + 8 = x - 8 + 8$ ← Rewrite the equation to show that 8 is added to both sides.

$35 = x$ ← This is the solution after adding 8 to both sides.

Check by using 35 in place of x.

$27 \stackrel{?}{=} 35 - 8$

$27 = 27$, so $x = 35$ is the correct solution.

Use $m + 17 = 43$ for Exercises 1–4.

1. What operation is shown in this equation? **addition**

2. What operation will you use to get m by itself? **subtraction**

3. Rewrite the equation showing subtracting from both sides of the equation. $m + 17 - 17 = 43 - 17$

4. What is the value of m? $m = 26$

Holt Mathematics

Puzzles, Twisters & Teasers
Clean Solutions!

Find the solution for each equation below. Write the letter of each variable on the line above the correct answer at the bottom of the page to solve the riddle.

1. $y - 35 = 17$ $y = 52$

2. $h - 40 = 26$ $h = 66$

3. $a + 16 = 43$ $a = 27$

4. $110 = e + 66$ $e = 44$

5. $97 = w - 44$ $w = 141$

6. $n - 8 = 3$ $n = 11$

7. $356 = g - 218$ $g = 574$

8. $652 + t = 800$ $t = 148$

9. $16 = o - 124$ $o = 140$

10. $63 + m = 903$ $m = 840$

11. $k + 18 = 98$ $k = 80$

12. $d - 27 = 54$ $d = 81$

13. $c - 50 = 23$ $c = 73$

14. $l - 62 = 937$ $l = 999$

Why did the robber take a shower?

H	E		W	A	N	T	E	D		T	O
66	44		141	27	11	148	44	81		148	140

M	A	K	E		A		C	L	E	A	N
840	27	80	44		27		73	999	44	27	11

G	E	T		A	W	A	Y
574	44	148		27	141	27	52

Holt Mathematics

Holt Mathematics

Solve.

1. $16 = n \div 2$

$n = 32$

2. $\frac{e}{10} = 8$

$e = 80$

3. $25 = \frac{x}{6}$

$x = 150$

4. $18 = \frac{d}{3}$

$d = 54$

5. $a \div 12 = 7$

$a = 84$

6. $30 = b \div 4$

$b = 120$

Solve and check.

7. $7w = 49$

$w = 7$

8. $75 = 3x$

$x = 25$

9. $60 = 12p$

$p = 5$

10. $77 = 11m$

$m = 7$

11. $4h = 48$

$h = 12$

12. $9y = 54$

$y = 6$

13. $2x = 30$

$x = 15$

14. $45 = 5s$

$s = 9$

15. $6z = 42$

$z = 7$

16. The Fruit Stand charges $0.50 each for navel oranges. Kareem paid $4.00 for a large bag of navel oranges. How many did he buy?

8 oranges

17. Jenny can type at a speed of 80 words per minute. It took her 20 minutes to type a report. How many words was the report?

1,600 words

18. At the local gas station, regular unleaded gasoline is priced at $1.10 per gallon. If it cost $16.50 to fill a car's gas tank, how many gallons of gasoline did the tank hold?

15 gallons of gasoline

Solve each equation. Check your answer.

1. $68 = \frac{r}{4}$

$r = 272$

2. $k \div 24 = 85$

$k = 2,040$

3. $255 = \frac{x}{4}$

$x = 1,020$

4. $42 = w \div 18$

$w = 756$

5. $\frac{a}{15} = 22$

$a = 330$

6. $82 = b \div 5$

$b = 410$

7. $\frac{c}{7} = 9$

$c = 63$

8. $28 = z \div 3$

$z = 84$

9. $\frac{y}{12} = 10$

$y = 120$

Solve each equation. Check your answer.

10. $52w = 364$

$w = 7$

11. $41x = 492$

$x = 12$

12. $410 = 82p$

$p = 5$

13. $35d = 735$

$d = 21$

14. $195 = 65h$

$h = 3$

15. $4k = 140$

$k = 35$

16. $110 = 5e$

$e = 22$

17. $27a = 216$

$a = 8$

18. $96 = 12n$

$n = 8$

19. Ashley earns $5.50 per hour babysitting. She wants to buy a CD player that costs $71.50, including tax. How many hours will she need to work to earn the money for the CD player?

13 hours

20. A cat can jump the height of up to 5 times the length of its tail. How high can a cat jump if its tail is 13 inches long?

$\frac{h}{5} = 13$; $h = 65$ inches

Solve each equation. Check your answer.

1. $765 = \frac{n}{12}$

$n = 9,180$

2. $\frac{m}{9} = 26$

$m = 234$

3. $\frac{a}{12} = 14$

$a = 168$

4. $3g = 165$

$g = 55$

5. $308 = 44b$

$b = 7$

6. $27e = 405$

$e = 15$

Translate each sentence into an equation. Then solve the equation.

7. The product of a number w and 145 is 725.

$145w = 725$; $w = 5$

8. The quotient of a number f and 21 is 14.

$f \div 21 = 14$; $f = 294$

9. A number b times 23 equals 253.

$23b = 253$; $b = 11$

10. A number k divided by 15 equals 47.

$k \div 15 = 47$; $k = 705$

11. A number n multiplied by 12 is 84.

$12n = 84$; $n = 7$

12. Ten divided into a number m equals 54.

$m \div 10 = 54$; $m = 540$

13. About three tons of ore must be mined and processed to produce a single ounce of gold. How many tons of ore are required to produce a pound of gold?

48 tons of ore

14. It costs about $2,931 per hour to operate a Boeing 757 airplane. Find the cost to operate a Boeing 757 during a 5-hour flight.

$14,655

15. Each person in the United States eats an average of 23 quarts of ice cream per year. At this rate, the seventh-grade class will eat about 1,794 quarts of ice cream this year. How many students are in the seventh-grade class?

78 students

16. Jumbo shrimp sell for $14.99 per pound in a local supermarket. One customer spent $74.95 on shrimp for a dinner party. How much shrimp did this customer purchase for the party?

5 pounds

When you solve an equation, you must get the variable by itself. Remember, what you do to one side of an equation, you must do to the other side.

• To solve a division equation, multiply both sides of the equation by the same number.

Solve and check: $\frac{a}{3} = 4$.

$$\frac{a}{3} = 4$$
$$\frac{3a}{3} = 1a = a$$
$$(3)\frac{a}{3} = 4(3)$$
$$a = 12$$

Check: $\frac{a}{3} = 4$

$$\frac{12}{3} \stackrel{?}{=} 4$$
$$4 \stackrel{?}{=} 4 \checkmark$$

Solve and check.

1. $\frac{x}{6} = 3$

$x = 18$

2. $\frac{s}{8} = 8$

$s = 64$

3. $\frac{c}{10} = 7$

$c = 70$

4. $\frac{n}{3} = 12$

$n = 36$

• To solve a multiplication equation, divide both sides of the equation by the same number.

Solve and check: $5k = 30$.

$$5k = 30$$
$$\frac{5k}{5} = 1k = k$$
$$\frac{5k}{5} = \frac{30}{5}$$
$$k = 6$$

Check: $5k = 30$

$$5(6) \stackrel{?}{=} 30$$
$$30 \stackrel{?}{=} 30 \checkmark$$

Solve and check.

5. $2w = 16$

$w = 8$

6. $4b = 24$

$b = 6$

7. $9z = 45$

$z = 5$

8. $10m = 40$

$m = 4$

Challenge
Shape Up

Find the value of each shape in Exercises 1–8. Then use the values to answer the questions below.

1. 

$2 \times \text{(butterfly)} = 32$
$4 \times \text{(beetle)} = \text{(butterfly)}$

$\text{(butterfly)} = \underline{16}; \text{(beetle)} = \underline{4}$

2. 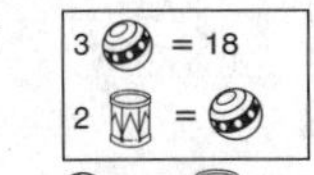

$3 \times \text{(drum)} = 18$
$2 \times \text{(drum)} = \text{(oval)}$

$\text{(drum)} = \underline{6}; \text{(oval)} = \underline{3}$

3. 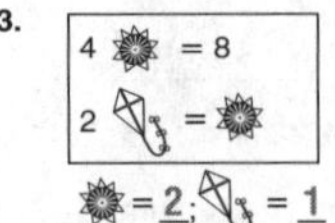

$4 \times \text{(sun)} = 8$
$2 \times \text{(kite)} = \text{(sun)}$

$\text{(sun)} = \underline{2}; \text{(kite)} = \underline{1}$

4. 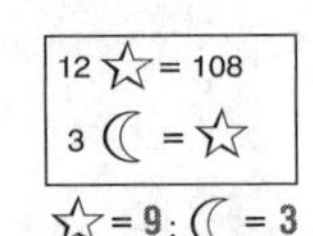

$12 \times \text{(star)} = 108$
$3 \times \text{(moon)} = \text{(star)}$

$\text{(star)} = \underline{9}; \text{(moon)} = \underline{3}$

5.

$24 \times \text{(shell)} = 192$
$32 \times \text{(shell)} = 0$

$\text{(shell)} = \underline{8}; \text{(shell)} = \underline{0}$

6. 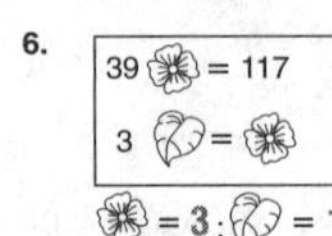

$39 \times \text{(leaf)} = 117$
$3 \times \text{(flower)} = \text{(leaf)}$

$\text{(leaf)} = \underline{3}; \text{(flower)} = \underline{1}$

7. 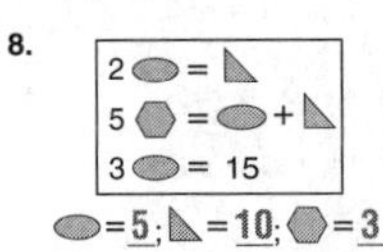

$4 \times \text{(apple)} = \text{(banana)} + \text{(pear)}$
$\text{(banana)} = \text{(pear)} + \text{(apple)}$
$2 \times \text{(banana)} = 30$

$\text{(banana)} = \underline{15}; \text{(apple)} = \underline{6}; \text{(pear)} = \underline{9}$

8.

$2 \times \text{(oval)} = \text{(triangle)}$
$5 \times \text{(hexagon)} = \text{(oval)} + \text{(triangle)}$
$3 \times \text{(oval)} = 15$

$\text{(oval)} = \underline{5}; \text{(triangle)} = \underline{10}; \text{(hexagon)} = \underline{3}$

9. What was the population of the United States in 1610?

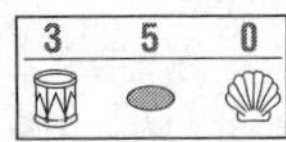

3 5 0

10. What was the population of the United States in 2000?

2 8 1, 4 2 1, 9 0 6

95
Holt Mathematics

Problem Solving
Solving Equations by Multiplying or Dividing

Write the correct answer.

1. The Panama Canal cost $387,000,000 to build. Each ship pays $34,000 to pass through the canal. How many ships had to pass through the canal to pay for the cost to build it?

11,383 ships

2. The rate of exchange for currency changes daily. One day you could get $25 for 3,302.75 Japanese yen. Write and solve a multiplication equation to find the number of yen per dollar on that day.

$25y = 3{,}302.75; y = 132.11;$

132.11 yen per dollar

3. Franklin D. Roosevelt was in office as president for 12 years. This is three times as long as Jimmy Carter was president. Write and solve an equation to show how long Jimmy Carter was president.

$12 = 3y; y = 4;$ **4 years**

4. The mileage from Dallas to Miami is 1,332 miles. To the nearest hour, how many hours would it take to drive from Dallas to Miami at an average speed of 55 mi/h?

24 hours

Choose the letter for the best answer.

5. The total bill for a bike rental for 8 hours was $38. How much per hour was the rental cost?

A $8 per hour
B $4.75 per hour
C $30 per hour
D $5.25 per hour

6. If a salesclerk earns $5.75 per hour, how many hours per week does she work to earn her weekly salary of $207?

F 30 hours
G 32 hours
H 36 hours
J 4 hours

7. At a cost of $0.07 per minute, which equation could you use to find out how many minutes you can talk for $3.15?

A $0.07 \div m = \$3.15$
B $\$3.15 \cdot m = \$0.07$
C $\$0.07m = \$3.15$
D $\$0.07 \div \$3.15 = m$

8. Which equation shows how to find a runner's distance if he ran a total of m miles in 36 minutes at an average of a mile every 7.2 minutes?

F $36 \div m = 7.2$
G $7.2 \div m = 36$
H $36m = 7.2$
J $7.2 \div 36 = m$

96
Holt Mathematics

Reading Strategies
Follow a Procedure

The opposite of multiplication is division: $12 \cdot 3 = 36$, and $36 \div 3 = 12$
The opposite of division is multiplication: $48 \div 12 = 4$, and $4 \cdot 12 = 48$
From these examples you can see that:
division "undoes" multiplication, and **multiplication "undoes" division.**
To solve multiplication and division equations:

- Get the variable by itself on one side of the equation.
- Keep the equation in balance by using the same operation on both sides.

Example:
$84 = 7x$ ← Get the variable by itself. This is a multiplication equation, so divide to "undo" the multiplication.

$\dfrac{84}{7} = \dfrac{7x}{7}$ ← Rewrite the equation to show that both sides are divided by 7.

$12 = x$ ← This is the solution after dividing both sides by 7.

Check using 12 in place of x:
$84 \overset{?}{=} 7(12)$
$84 = 84$, so $x = 12$ is the solution.

Example:
$\dfrac{m}{15} = 8$ ← Get the variable by itself. Multiply to "undo" division.

$\dfrac{m}{15} \cdot 15 = 8 \cdot 15$ ← Rewrite the equation to show that both sides are multiplied by 15.

$m = 120$ ← This is the solution after multiplying both sides by 15.

Check by using 120 in place of m.
$\dfrac{120}{15} \overset{?}{=} 8$
$8 = 8$, so $m = 120$ is the solution.

Use $108 = 9y$ for Exercises 1–3.

1. What operation will you use to solve the equation?

division

2. Rewrite the equation using the inverse operation on both sides.

$\dfrac{108}{9} = \dfrac{9y}{9}$

3. What is the value of y?

$y = 12$

97
Holt Mathematics

Puzzles, Twisters & Teasers
Doctor! Doctor!

Find the solution for each equation below. Write the letter of each variable on the line above the correct answer at the bottom of the page to solve the riddle.

1. $a \div 25 = 4$ $a = 100$
2. $w \times 18 = 18$ $w = 1$
3. $r \div 8 = 5$ $r = 40$
4. $3h = 96$ $h = 32$
5. $72 = 8d$ $d = 9$
6. $12 = y \div 4$ $y = 48$
7. $17 = n \div 8$ $n = 136$
8. $85 = 17o$ $o = 5$
9. $3e = 63$ $e = 21$
10. $9 = u \div 3$ $u = 27$
11. $6b = 222$ $b = 37$
12. $7m = 84$ $m = 12$
13. $150 = 3t$ $t = 50$
14. $9s = 99$ $s = 11$

Patient: Doctor! Doctor! I feel like an umbrella!

Doctor:

W	H	Y	,		Y	O	U
1	32	48			48	5	27

M	U	S	T		B	E
12	27	11	50		37	21

U	N	D	E	R		T	H	E
27	136	9	21	40		50	32	21

W	E	A	T	H	E	R
1	21	100	50	32	21	40

98
Holt Mathematics

122
Holt Mathematics

Holt Mathematics